SpringerBriefs in Applied Sciences and Technology

Computational Intelligence

W0037185

Series Editor

Janusz Kacprzyk, Systems Research Institute, Polish Academy of Sciences, Warsaw, Poland

SpringerBriefs in Computational Intelligence are a series of slim high-quality publications encompassing the entire spectrum of Computational Intelligence. Featuring compact volumes of 50 to 125 pages (approximately 20,000-45,000 words), Briefs are shorter than a conventional book but longer than a journal article. Thus Briefs serve as timely, concise tools for students, researchers, and professionals.

More information about this subseries at http://www.springer.com/series/10618

Siba Kumar Udgata · Nagender Kumar Suryadevara

Internet of Things and Sensor Network for COVID-19

 Springer

Siba Kumar Udgata
School of Computer and Information
Science
University of Hyderabad
Hyderabad, India

Nagender Kumar Suryadevara
School of Computer and Information
Sciences
University of Hyderabad
Hyderabad, India

ISSN 2191-530X   ISSN 2191-5318   (electronic)
SpringerBriefs in Applied Sciences and Technology
ISSN 2625-3704   ISSN 2625-3712   (electronic)
SpringerBriefs in Computational Intelligence
ISBN 978-981-15-7656-0   ISBN 978-981-15-7654-6   (eBook)
https://doi.org/10.1007/978-981-15-7654-6

This Springer imprint is published by the registered company Springer Nature Singapore Pte Ltd.
The registered company address is: 152 Beach Road, #21-01/04 Gateway East, Singapore 189721,
Singapore

Preface

The unfortunate outbreak of COVID-19 has come as a shock to society and health systems. The COVID-19 pandemic has changed the discourse and shaken the existing system and surpassed provincial, radical, conceptual, spiritual, social, and pedagogical boundaries. In the present pandemic situation, researchers, scientists, engineers, and technocrats cutting across national boundaries are working hard to combat the challenges owing to COVID-19. They are constantly looking for a practical and cost-effective solution to mitigate the challenges. There has been a lot of work over the past decade in the development of sensor technology and IoT framework for the betterment of society.

In this book, we make an honest and sincere attempt to present the advances in sensor technology and IoT on the theme "IoT and Sensor Network for Covid-19." Governments across the world and healthcare specialists are striving to come out with guidelines and develop systems for managing the COVID-19 pandemic. The COVID-19 management is mainly based on (1) maintaining proper physical hygiene, (2) enforcing social distancing, (3) protecting face (use of mask), (4) avoiding touching of the face, (5) contactless control of appliances to stop spreading the virus, (6) self-isolation in case of suspicion, (7) contact tracing process, (8) home quarantine for patients, (9) screening and monitoring of individuals, and (10) mass screening and testing. Different chapters in the book highlight the COVID-19 pandemic challenges faced by individuals and healthcare systems in utilizing the benefits of the latest technologies, such as the sensor network and the IoT.

The book evokes the framework and numerous capabilities of IoT in a clear and easy-to-understand way. We have attempted to cover all the vital information identified with the IoT framework and how it is helping the healthcare provider, governments, NGOs, industries, and the general public to deal with the crisis. This book presents the investigation, outcomes, and applications coming out from the utilization of the IoT innovation. The objectives of the book are (i) to assemble and integrate information on the status of the present and most recent uses of IoT in the healthcare services, particularly for mitigating the COVID-19 crisis, and (ii) to furnish the intricacies to the reader pertaining to the design, development, and integrating hardware and software components with reasonable comprehension and the specific thought of IoT. Besides the above objectives, we also believe that this book will ignite some young and creative minds to come up with innovative solutions for COVID-19 and similar crises.

The book is organized as follows:

Chapter 1 gives an overview of the COVID-19 challenges and the issues society is facing due to this pandemic. It describes the various challenges involving the use of technology, data, and also ethical and moral challenges. The chapter also briefly introduces the different domains in which the technology is being used. Chapter 2 presents a brief introduction of IoT architecture, wireless communication standards, different sensors, and microcontrollers. This chapter presents the required fundamentals of the technologies required for basic understanding and development of IoT-based devices and systems. In Chap. 3, we present the COVID-19 crisis, the sensor technology, and the Internet of Medical Things (IoMT). We present a detailed description of the available technology, sensing system used for healthcare in the view of COVID-19 crises. Chapter 4 presents a detailed description of the popular and available IoT devices and framework that is being used worldwide for managing the COVID-19 crisis. This chapter includes a detailed description of recent research being carried out at different universities, research institutes, and industries to help the government, individuals, and society at large. Finally, in Chap. 5, we discuss the possible new applications based on the future advances in the IoT and sensor network like 5G/6G communication and Meshstatic.

The book will be useful for an IoT enthusiast, developer, or for someone curious about the Internet of things. A basic understanding of electronic hardware, networking, and basic programming skills would do wonders in implementing the ideas discussed in this book.

Hyderabad, India   Prof. Siba Kumar Udgata
                                                  Dr. Nagender Kumar Suryadevara

Contents

1   **COVID-19: Challenges and Advisory**   1
    Introduction   1
    Technologies and Their Applications for Fighting the COVID-19   3
    Challenges and Opportunities   3
        Challenges Involving Data Collection.......................   4
        Information Reliability Challenge for Data Collection   5
        Challenge of Information Modality   6
        Scalability of Edge-Based Artificial Intelligence (AI) Model
        for COVID-19 Data Processing   7
        Challenge Location Information Privacy......................   7
        Challenge of Data Visualization and Presentation   8
        Human Factor Challenge   8
    Recommended Precautions by Health Organizations   9
        Safety Measures   10
        Contact Tracing   11
        Safe Blues: A Method for Estimation and Control of the Spread
        in the Fight Against COVID-19   13
        Advisory for the Use of Sensors and IoT Framework   14
    References   15

2   **IoT and Sensor Network**...................................   19
    Introduction   19
    Layers of IoT Architecture   20
    Sensors Connectivity   21
    Gateway/Network Layer   22
    Management Service Layer   23
    Application Layer...   24
    IoT Protocols   24
        Link-Layer Protocols....................................   24

Network/Internet Layer Protocols   25
Transport Layer Protocols   26
Application Layer Protocols.   26
IoT Programming Languages and Coding   27
IoT Implementation and General Usage   31
References   36

3   COVID-19, Sensors, and Internet of Medical Things (IoMT)   39
Introduction   39
Sensor Systems for COVID-19   40
Disease Tracking Through Crowdsensing   42
UAV-Based Health Surveillance and Alerting.   44
Ongoing IoT Research for Addressing COVID-19
Challenges   45
Effect of COVID-19 on IoT Activities   50
References   52

4   Advances in Sensor Technology and IoT Framework to Mitigate
COVID-19 Challenges.   55
Introduction   55
System for Avoiding Face Touch   55
The Design Approach of the "Face Touch" Detection System
Include   56
An Intelligent IoT Framework for Gesture Recognition to Achieve
Contactless Interface and Social Distance.   57
Hardware and Connections Used   59
Implementation   62
The Interface Between MPU6050 and ESP8266   63
Smart Glasses for Monitoring and Controlling the Infection Spread   64
The Methodology   66
Smart Wearable Thermometer for Continuous Temperature
Monitoring.   67
Connected Thermometers for Large-Scale Temperature
Screening   68
Small and User-Friendly Contactless Thermometer   69
IoT Buttons for Generating Alert for Maintaining Cleanliness   69
Smart Helmet-Based Novel COVID-19 Detection and Diagnosis   70
IoT-Based Drone Technology for Mitigating Challenges Owing
to Coronavirus   73
IoT-Based Wearable Band to Track COVID-19 Quarantine
Patients   75
COVID-19 Intelligent Diagnosis and Treatment Assistant Program
(NCapp).   77
Robots and Drones in Service of the Society During COVID-19   78

Systems That Indirectly Ensures Social Distancing 78
   Monitoring Food Safety During the COVID-19 Pandemic 79
   IoT Devices and Modern Dining Experiences 80
   Inventory Control and Management 80
References 81

5 Future Possible Applications 83
Progressions Related to IoT for COVID-19 83
Global Innovative Progressions to Determine COVID-19 Cases
Quickly 85
Significant Uses of IoT for COVID-19 Pandemic 85
Various Issues and Future Extent of the Investigation 87
Effect on the IoT Market Will be Blended 88
IoT Technological Advancements for Healthcare
Applications 89
   Meshtastic 89
5G/6G Communications 90
Smart Sensing: Improving the Capabilities of Sensors
Based on AI Technologies 91
References 91

Conclusion 93

About the Authors

Dr. Siba Kumar Udgata is currently a Professor of Computer and Information Sciences at the University of Hyderabad, India, where he directs a research group focusing on sensor networks, IoT, wireless communications, and intelligent algorithms. He worked as a Research Fellow at the United Nations University/International Institute of Software Technology (UNU/IIST), Macau. He has published more than 100 research papers in peer-reviewed journals and at international conferences. He has edited ten international conference proceedings for Springer LNAI, AISC and SIST. He is a recipient of the IBM SUR (Shared University Research) award for the project "Mobile Sensor network based rescue management system". He has successfully completed seven Government of India sponsored research projects in the domain of sensor network, IoT, and cognitive radio network.

Dr. Nagender Kumar Suryadevara received his Ph.D. degree from the School of Engineering and Advanced Technology, Massey University, New Zealand, in 2014. He has authored/co-authored two books and over 45 papers in various international journals, conferences, and book chapters. His research interests include wireless sensor networks, the Internet of things, and time-series data mining.

Chapter 1
COVID-19: Challenges and Advisory

Introduction

During the COVID-19 pandemic, governments and healthcare organizations must take some steps to reduce the spread of the virus and to treat the already infected persons. Infected patients with symptoms need to be hospitalized. Also, the patients without symptoms may still be releasing the virus, and therefore, they need to be quarantined for a predefined period (typically 14 days) until the virus release ends. The rapid increase of COVID-19 cases is putting immense pressure on the healthcare infrastructure.

Like other viruses, COVID-19 infection also goes through different stages such as

1. incubation in the host without symptoms,
2. the onset of full-blown symptoms,
3. a period during which viruses can still be released, and
4. finally, the recovered stage with no virus release.

Tracking down any "patient zero" is time-consuming and also resource-consuming, so also quarantining patients. The number of infected patients is increasing more rapidly than hospitals can afford to add beds, ventilators, monitors, personal protective equipment (PPE), and healthcare workers. Many of the hospital staff have also become infected themselves, generating even more stress on the remaining staff.

Researchers in physical sciences and engineering are engaged in research and innovation to take such challenges, describe new study problems, grow new theories, and generate user-centric explanations. In this chapter, an attempt has been made to provide a brief description of many innovative technologies involving sensors, wireless communication, the Internet of things (IoT), intelligent machine learning algorithms, and significant applications for the COVID-19 pandemic.

© The Editor(s) (if applicable) and The Author(s), under exclusive license to Springer Nature Singapore Pte Ltd. 2021
S. K. Udgata and N. K. Suryadevara, *Internet of Things and Sensor Network for COVID-19*, SpringerBriefs in Computational Intelligence,
https://doi.org/10.1007/978-981-15-7654-6_1

Throughout these substantial changes achieved by the COVID-19 pandemic, a few chances and difficulties have risen, especially in China. The Internet of Medical Things (IoMT) holds tremendous potential for applications in medicinal services on account of the social distancing and separation required to slow the pandemic, and also governments are continuously looking for endless ways to deal with human services [1].

Without a doubt, the problem associated with routine visits of the clinics and hospitals, the much feared and dreaded view of emergency clinics during the pandemic constrained the general population to look for different methods for access to human services. These are handled by online interviews offered by advanced smartphones and online clinical staff. The new innovative models like the Internet of Medical Things (IoMT) and computerized wellbeing are proposed that can change the approach in different countries and facilitates the best means to control advanced wellbeing during both emergency and optional clinical intercessions. Significantly, the possibilities for constant determination, avoidance, and mediation are tested by the need to approve the computerized wellbeing approaches in manners that are experimentally strong and socially justifiable [1].

Internet (Web) of Medical Things (IoMT) is assuming an essential job in the social insurance industry to expand the precision, uncompromising quality, and efficiency of electronic gadgets. Analysts contribute to an automated social insurance framework by interconnecting the accessible clinical assets and human services administrations. As the Internet of things (IoT) merges different spaces, IoMT has emerged into IoT for various healthcare services [2]. The current worldwide test of the COVID-19 pandemic has outperformed the conventional, radical, calculated, profound, social, and educational limits. Internet of things (IoT) empowered human services framework is valuable for the legitimate checking of COVID-19 patients by utilizing an interconnected system. This innovation assists with expanding the fulfillment at remote locations and lessens the readmission rate in the clinic. IoT usage impacts on decreasing medicinal services cost and improves treatment results of the infected patient. Accordingly, current explorations are endeavored to investigate, diagnose, discuss, and feature the general utilizations of the all-around demonstrated IoT reasoning by offering a comprehensive guide to handle the COVID-19 pandemic [3].

There is a minimum prerequisite for scholastic exploration on a few parts of this exceptionally infectious illness to discover successful methods for control and treatment of the disease, in the present time being, and also for the future. We try to explore a few open doors for scholarly exploration identified with COVID-19 and have additionally identified suitable recommendations to contain, forestall, and treat this viral contamination [4]. Cognitive Internet of Medical Things (CIoMT) is a disruptive innovation that gives constant clinical information about COVID. CIoMT is a promising innovation for quick finding, observing, remote wellbeing, and dynamic following for control of the disease. It helps in contact following and grouping, screening and surveillance, and in decreasing the remaining task of the clinical examination. In the period of cutting-edge computerized innovation, many human services dependent on the Internet of things (IoT) are picking up significance to manage the current COVID-19 pandemic. The CIoMT is investigated to handle the comprehensive test. This idea of CIoMT is most appropriate to this pandemic as each individual is to be associated and checked through an extensive system that requires a range of healthcare executives [5].

Technologies and Their Applications for Fighting the COVID-19

Artificial intelligence (AI) is a creative innovation that is useful to battle the COVID-19 pandemic. This innovation is helpful for appropriate screening, following, and foreseeing the current and future patients. The significant uses of this AI are for new location and identification of the disease. Artificial intelligence is used to improve immunizations and medications to reduce the remaining task at hand of medicinal service staff [6].

COVID-19 (Coronavirus) pandemic has created a demand for fundamental social insurance gear, drugs alongside the necessity for advanced data-centric applications. Industry 4.0 is the fourth electro-mechanical transformation, which can satisfy modified prerequisites during the COVID-19 emergency. This disruptive change has begun with the use of advanced assembling and computerized data innovations [7].

The setting of whole urban areas in "lockdown" straightforwardly influences urban economies on multiple fronts and at different levels, including both social and financial viewpoints. This is being underlined as the scenario that makes strides in different nations, driving toward a worldwide wellbeing crisis, and global cooperation is looked for in various regions. The urban point of view and advances on how smart city systems should progress in improving normalization conventions for expanded information gathering and analysis in case of outbreaks like the current pandemic [8].

The clinical environment idea for battling coronavirus sickness (COVID-19), consolidated usage of purpose-of-care (POC) diagnostics, and the Internet of Medical Things (IoMT) [9] is shown in Fig. 1.1.

Challenges and Opportunities

The current flurry of innovations is trying to develop devices, modules, and systems to help specialists forestall the additional spread of COVID-19 while also helping those that are tragically contaminated. IoT, specifically and particularly when joined with other transformative advancements like cloud and AI, has great use in a broad scope of utilization during the emergency.

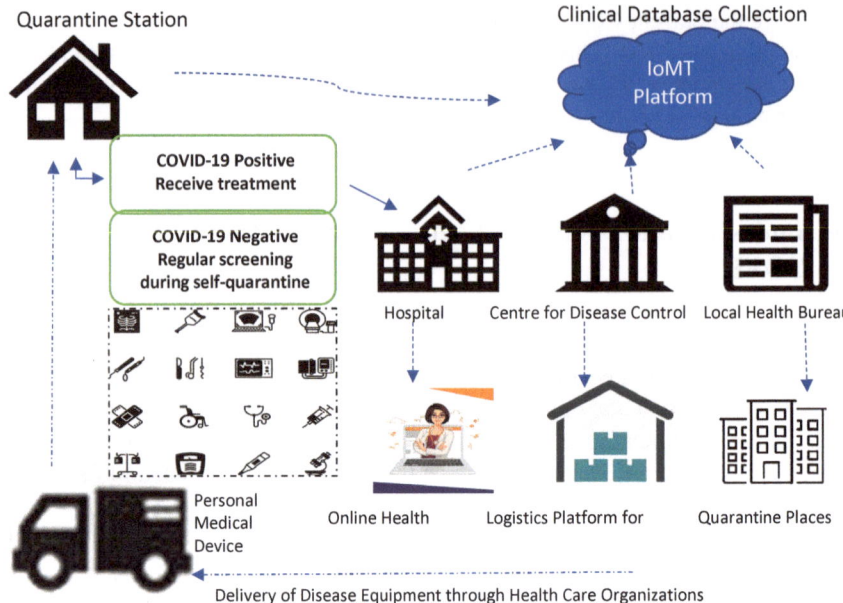

Fig. 1.1 The healthcare management system during a pandemic [9]

The broad scope of utilization demonstrates that IoT is an appropriate response in unraveling one of the novel difficulties created by COVID-19. The use of technology about IoT poses particular challenges and opportunities. The following are the significant challenges and opportunities in the present pandemic while considering the available technologies augmented with the Internet.

Challenges Involving Data Collection

During the beginning of widespread sickness outbreaks like COVID-19, the essential target of a COVID-19 framework is to gather data from the population. Few challenges have been identified to find and acquire the information identified with the COVID-19 spread. For example, while leading basic technological phrases such as "COVID-19 technology," many ventures are trying to obtain internet-based health information. This may lead to the gathering of other undesirable significant information beyond human recognition or body temperature or places of visit.

There is a possibility that the agency which is collecting data may misuse it. Even other agencies can misuse the data through the collecting agency. In this manner, getting an assortment of applicable online networking information that coordinates to the correct arrangement of data becomes difficult. An extraordinary segment of online networking information may, in the long run, end up being excess (e.g., retweets) or reworded from a separate unique post.

Moreover, a decent portion of Web-based health information is seen to be transient and also transitory. For instance, individuals may erase their past posts and online stores (i.e., Twitter and Facebook servers) for undisclosed and unknown reasons. Notwithstanding that, internet-based APIs, for example, Twitter, regularly force different rate confinements, which can intensely block the information aggregation during an emergency and critical situation. The information assortment process for COVID-19 accordingly requires a device that can find, get, and store the applicable data from clients continuously across Web-based health stations securely and reliably.

Information Reliability Challenge for Data Collection

Information produced by the obscure human sources on Web-based social networking lacks information reliability (i.e., the trustworthiness of the collected data). One significant challenge in dealing with online health for IoT innovation is to separate reliable data from temperamental human sources with obscure source questionable quality. Some research work tried to address and moderate the information quality issue. For example, Wang et al. [10, 11] introduced a structure to gauge the dependability of information sources and the correctness of the detailed estimations in internet-based health posts, which comes closer to the estimation hypothesis. Zhang et al. [12, 13] used past systems' experience to address the adaptability and physical limitations and utilized the improved plans to verifiable social detecting applications. Yin et al. [14] presented a model "Truth Finder," a probabilistic calculation that uses iterative weight updates to improve the nature and correctness of the information in social detecting. While many endeavors have been made while creating robust social detecting arrangements, certain restrictions affect these arrangements from being applied to COVID-19 situations. One drawback of current social detecting plans is that they exclusively depend on viral online life information. There are no other methods for checking the trustworthiness of the data during the COVID-19 parlance. Existing techniques are not customized toward disease discovery, which may prompt a forecast of faulty instances of COVID-19. For example, an individual posting an issue of breathing trouble may not experience the ill effects of COVID-19. It might be required to break down the patient's different qualities and profiles, depending on prior posts. Thus, it remains an uncertain test to create dependable social detecting models that can investigate the vulnerability in the information.

Challenge of Information Modality

While information assortment is a diagnostic test in utilizing social detecting for following the COVID-19 spread, a more noteworthy problem exists in preparing the hurriedly created signs consisting of vast numbers of measurements. This test is distinguished as an information-oriented methodology in social detecting where a lot of unfiltered and unstructured information with different modalities requires to be processed [15–17]. In particular, information-based methodology alludes to the information pervasive in online networking, such as text, picture, location, audio, and video. Also, each type can additionally include distinctive dimensionality, which makes the information-based methodology considerably challenging. Instances of dimensionality can extend along with reports of (i) closeness to contaminated areas, (ii) the number of suspected cases, (iii) number and kinds of indications, (iv) intensity of side effects (i.e., low, moderate, or acute), (v) recuperation rate, (vi) death rate and (vii) the number of self-isolated cases. Ongoing social detecting devices primarily focus on examining the content information in internet-based life. This pattern is supported by the way that image and video information handling require substantial computational overhead.

Therefore, existing techniques do not concentrate on aggregating and dealing with different kinds of information, which may create false identification of COVID-19 infected persons. For instance, an individual may tweet about having COVID-19, yet dependent on a picture posted with the tweet; it might turn out that the individual's indications have come due to a hypersensitive response. Aggregating text with other information, such as picture and location information, may yield increasingly precise forecasts of the COVID-19 spread. With these principles, given the sheer volumes of multi-modular information produced by the Web-based social networking clients about the COVID-19 scenario, arrangements should be created to use the diverse methodology of information productively. Additionally, since multi-modular information processing characteristically requires more processing power, care must be taken to strike a trade-off between location exactness and the computational requirement. Many challenges evolving from the information-based methodology are: (i) how to proficiently intertwine the various kinds of Web-based health information identified with COVID-19 into one information stream? (ii) how do we structure computations, to process a wide assortment of social information for the exact estimation of the COVID-19 spread? (iii) how to accelerate the examination of multimodular information for quicker COVID-19 spread recognition by appropriating the calculation over different devices?

Scalability of Edge-Based Artificial Intelligence (AI) Model for COVID-19 Data Processing

Because of the worldwide COVID-19 outbreak, it is imperative to use and depend on versatile AI-based strategies that can adequately screen the condition of the spread from the social detecting information over any locality continuously. This requires the adaptable AI methodology that can be promptly deployed over the edge devices (e.g., smartphones, IoT gadgets, drones) to reduce idleness and data transfer capacity utilization and yield quicker data extraction for the COVID-19 spread. Unfortunately, existing AI-based methods, like deep neural network (DNNs), Multi-Layer Perceptron (MLPs), Hopfield network, Bayesian network, and recurrent neural networks (RNNs) have been initially used for high power computing platforms like GPU groups and are not customized for low computing power devices operating at the edge of the system [18–21]. Specifically, current AI models are related to periodic model updates that work in an integrated manner, which requires a large system transmission capacity. Notwithstanding that standard AI models need to set and refresh the model boundaries before generating reliable forecasts. Hence, regardless of whether the current AI methods could be adapted to execute on the edge devices, because of their substantial computation requirements for the model preparation, they would deplete the energy of the versatile and energy-constrained edge devices quickly and reduce the lifetime of the device. A couple of challenges associated with the AI model would be (i) how to parallelize the AI model preparation process over the edge devices to accelerate the model preparation and enhance data transfer capacity? (ii) how to advance the AI computations to run productively on the low powered edge equipment? (iii) how to modularize the AI computations so that they can be effectively deployed over a large number of edge devices?

Challenge Location Information Privacy

One compelling issue in social detecting is client security, whereby the individual data of the online clients falls into inappropriate hands. Location-area information shared by clients can likewise be utilized to uncover other private data (e.g., ethnicity, race, monetary status). Online networking clients do not commonly agree to share and are also not required by applications. Many concerns have been raised across the country boundaries regarding this issue. The Government of India sponsored and recommended the "Arogyasetu" application. Similar applications worldwide are attracting criticism from a section of users and intellectuals regarding the privacy of data and possible misuse and state-sponsored surveillance. Along these lines, it has been seen that because of the worry of one's area and privacy, numerous internet-based clients tend not to share their area data while revealing their perceptions in the Web-based model. It is found that under 10% of the tweets are geo-labeled (i.e., contained a topographical location area of the clients). Accordingly, applications that vigorously depend on the area metadata from the Web-based social networking present to indicate the COVID-19 spread may fail to meet expectations when the quantity of geo-labeled internet-based data is quite scant. Thus, to follow up on the advancement of the COVID-19 spread, it is essential to acquire the specific areas of the spread of infected persons, i.e., COVID-19 affected areas [22–24].

Challenge of Data Visualization and Presentation

With the rapidly advancing conditions during the COVID-19 outbreak in different regions, it has become mandatory to introduce and present the data of the infection spread to the citizens in an ideal manner. This requires a data introduction framework that can both present information on the infection spread progressively and also alert individuals on time. A few strategies have been executed to introduce infection spread updates to the public in the recent past through various intuitive sites. Such a strategy for data dissemination and assortment exclusively depends on aggregating information from different news portal Really Simple Syndication (RSS) feed and the data sites which can prompt infections, recoveries, deceased, trend, etc. about the latest phenomena. Because of their distributed nature of data slithering and examining, existing electronic procedures cannot be straightforwardly applied to social detecting, which envelops unstructured and multi-dimensional social information.

Notwithstanding that, sites and smartphone applications depend on the steady accessibility of both the Internet and edge nodes, both of which may not be possible in all conditions. In this way, crucial data may not reach everybody, particularly with the old and less technically knowledgeable people without access to personal computers or smartphones. Due to the multi-lingual, multi-cultural, and varied educational and financial background, the presentation of data to the general population in an effective manner remains a significant challenge. Natural language mode and graphical displays have been widely used for data visualization and presentation but still is far from adequate and reachable to all sections of the population [25].

Human Factor Challenge

In any society, technology adaptation by the general public remains a big challenge. This issue becomes critical in an emergency like the COVID-19 pandemic. The human factor is a big challenge in data collection and aggregation. Given the fearful concerns and anxiety among the overall population during the COVID-19 pandemic, individuals are excessively enthusiastic, exciting, or one-sided in communicating their sentiments in the Web-based portals and other crowdsensing applications. Such unusual enthusiasm and conduct can conceivably trigger distorted or misconstrued perceptions and subsequently yield unwanted outcomes. Thus, a primary challenge would be how to deal with the mindset of the people, while containing the worry at all levels? It is also essential to consider the individual segment more closely and model how individuals respond to the data introduced to them. A few people may end up being too sensitive, so care must be taken not to build up the justification for unwarranted turmoil. For instance, during the Ebola outbreak in Liberia in the year 2014, riots broke out among the people when authorities raised cautions of the outbreak. On the other side, we also have to understand that a good percentage of the populace tends to be indifferent to the conditions, disregard admonitions, and remain unusually quiet. The challenge is to find some kind of balance between presenting information and alerts so that people are informed enough to take precautions and prepare for eventuality and not be the reason for spreading panic among the citizens to avoid chaos. Thus, it stays a fundamental challenge to develop a successful and reliable model that can distinguish and detach the deception spread to produce dependable social signs showing the COVID-19 spread [26, 27].

Recommended Precautions by Health Organizations

You can decrease your chance of being contaminated or becoming an instrument for spreading COVID-19 by staying inside the home and safe. Some of the usual guidelines include

1. Routinely and thoroughly clean hands with an alcohol-based hand sanitizer and wash them with soap and water for about 20 s. Washing your hands with soap and water or using alcohol-based sanitizers deactivates/neutralizes infections that might present on your hands.
2. Ensuring a distance of 1–2 m (3–6 feet) separation between yourself and other people is the key. When somebody with COVID-19 infection talks, hacks, or wheezes, they shower little fluid beads from their nose or mouth, which may contain the virus that is supposed to be in the air for around 1–2 m before falling. Thus, there is every chance that if you are excessively close, you can get exposed to the beads, which contains the COVID-19 infection. One of the essential measures to follow is to abstain from going to crowded places. Where individuals meet up in groups, you are bound to come into close contact with somebody that has possible COIVD-19 infection. It is progressively difficult to keep up physical separation of 1–2 m (3–6 feet) always.
3. Wear a mask to cover the face, mostly nose, and mouth, to avoid entering the aerosol beads. This will also help you not to become an instrument to spread the infection in a case; unfortunately, you are infected and not even aware of it. We must ensure that all of us follow great respiratory cleanliness. This implies covering mouth and nose with your bowed elbow or tissue when you hack or sniffle. At that point, immediately discard the mask and wash hands. By following proper respiratory cleanliness, you shield the individuals around you from infections.

4. Avoid touching the face (nose, mouth, and eyes) to avoid infection. The most common source of infection is different surfaces, and through the surface, it infects the hands through inadvertent touch. Coronavirus can survive for quite a while on different surfaces, which is a significant reason behind its transmission. This infection can sully on various metal surfaces and remain active on these surfaces from hours to days, with a most extreme range on plastic and least on the copper surface. The liquor-based disinfectants can necessarily decrease the endurance of the infection. The two significant coronaviruses (SARS-CoV-2 and SARS-CoV-1) have critical continuing time on various metal surfaces, and their conduct is practically comparable on different metal surfaces and in mist concentrates. Thus, your hands become a source of infection, and touching your face is quite critical in getting infected [28].

5. Continuous monitoring of body temperature is also an adequate precaution for controlling the spread and seeking medical help. Increased body temperature has been one of the critical symptoms of COVID-19 infection. Although increased body temperature is not necessarily a confirmatory test of infection, it is one of the essential measurable parameters.

6. Home quarantine and self-isolation are preferable in case of suspected symptoms of COVID-19 to take care of yourself as well as your loved ones. Remain at home and self-detach even with minor side effects, such as hack, migraine, and mellow fever, until you recuperate. Evading contact with others shall shield them from conceivable COVID-19 and different infections. In case you have a fever, hack, and trouble breathing, look for clinical consideration, telemedicine, on-call medical assistance. National and nearby specialists shall have the most cutting-edge data on the circumstance in your general vicinity. This will secure you and help forestall the spread of infections and different diseases.

Sensor-oriented technology and IoT platforms have been leveraged to a greater extent to address the above precautions and used as a tool for enforcing people to follow these precautions.

Safety Measures

The world is now struggling to control the unprecedented spread of COVID-19 infections, which leads to a record number of morbidities and mortalities. Since there is no specific treatment for coronaviruses, and no foolproof method to contain the spread, there is an urgent need for global surveillance of individuals with active COVID-19 infection. An integrated digital disease surveillance system with the help of IoT is crucial to controlling this disease. Application of IoT in infectious disease epidemiology is an emerging field. The ubiquitous availability of smart technologies and increased risks of infectious disease spread through the globalization and interconnectedness of the world necessitates its use for preventing, predicting, and controlling emerging infectious diseases. Several countries are working on Web-based surveillance tools and epidemic intelligence methods to facilitate risk assessment and timely outbreak detection. However, the widespread use of the available technologies is still not put to practice to its full potential.

Contact Tracing

The Internet of things utilizes the interconnected network for the data-flow and exchange of data. It also enables the patients, civilians, and health administrators to connect with the service providers for suitable actions. By employing the different IoT model in the COVID-19 pandemic, the effective and efficient tracing of the patients and the suspicious cases can be reasonably assured. A few practical smartphone-based applications are developed for the benefit of the people. In this regard, the Government of India has launched a smartphone application named "Arogya Setu."

Similarly, the mobile application called "Close-Contact (English name of the Chinese app)" is launched for its citizens in China. These applications inform the app user about the closeness to the corona-positive person. The US Government, Singapore Government, Australian Government, and other countries also launched a similar mobile application for its citizens. Apart from COVID-19, it can alert and track similar diseases to improve the general public's safety. It digitally captures the patient's data and information through the designed mobile application without any human interaction.

Contact tracing is an observation and regulation technique proposed around 80 years back to control syphilis. Rather than overseeing individual cases while looking for clinical consideration, contact tracing attempts to detect the path of infection spread from infected patients to those they have been in close physical contact. A few methodologies for contact tracing are first-request, single-step, iterative, and review or retrospective was utilized previously during different outbreaks. The first-request method recognizes those individuals the patient quickly came into contact with and informs them concerning the potential infection and the need to look for clinical counseling or self-isolation. It does not bother about following the contacts of contacts, leaving that to second-request procedure as and when the initial request contact looks for clinical consideration.

Single-step contact tracing recognizes all individuals that the infected individual came into contact within the recent past. As any of those are additionally distinguished as contaminated, their contacts are detected, and the procedure follows. One issue with single-step contact tracing is that asymptomatic persons can spread the disease until they are identified and separated. However, iterative contact tracing keeps on following and reapply the significant indicative test to contacts iteratively before their condition may be identified through manifestation screening. The procedure proceeds until no newer infected persons are found. Retrospective contact tracing follows a similar process as either single-step or iterative with the expansion. It also works backward by considering the individuals with which the infected patient had

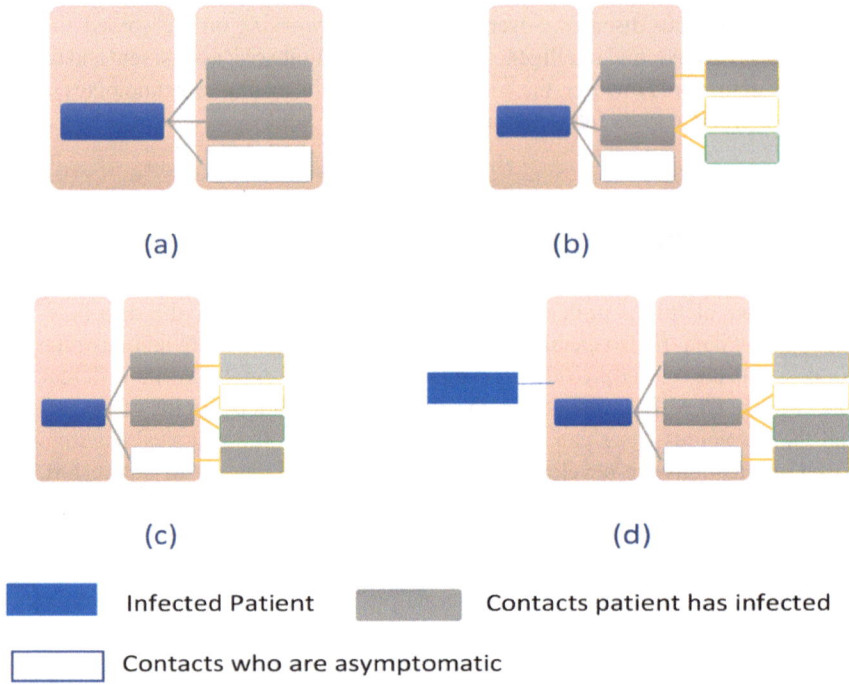

(a)   (b)

(c)   (d)

Infected Patient   Contacts patient has infected

Contacts who are asymptomatic

Fig. 1.2 a First order; **b** single-step; **c** iterative; and **d** retrospective contact tracing

been in contact within their ongoing past, intending to determine who it was that contaminated the patient. These methodologies are shown in the following Fig. 1.2.

With our expanded global village concept, worldwide populace, global aircraft travel, megacities, and mass travel, traditional contact following alone cannot contain the virus spread. Conventional contact following was utilized right during the SARS period. It ignored the possible contamination, which immediately spread through the more extensive network. Specialists and researchers stress upon some new methodologies that are required to do the contact tracing in the present scenario. Current contact following methods has been proposed utilizing universal and almost omnipresent smartphones. The smartphone advances facilitate recording and reporting when we have come into close physical contact with others. It is accepted that mechanized contact following will help in handling circumstances when we either do not know about or cannot remember, each contacted individuals. It gives (a) development centered portable programmed contact recording; (b) contact distinguishing proof, (c) contact notice, and (d) narrowcast information. This new technique leads to contact tracing applications like "COVIDSafe," "Arogy setu," etc., which will fundamentally lessen the probability of contamination of your loved ones and keep your family protected from COVID-19 infections.

IoT-based smart disease surveillance systems have the potential to be a significant breakthrough in the efforts to control the spread of the current pandemic in the current global situation. As much of the required infrastructure is already in place (i.e., smartphones, wearable technologies, internet access), this technology can have a significant impact on limiting the spread of the epidemic that involves only the collection and analysis of data already gathered. IoT and related new technologies are currently helping early recognition of spread and prevent the spread of infectious diseases that include the COVID-19. Smart disease surveillance systems based on IoT are quite affordable and would provide simultaneous reporting and monitoring, end-to-end connectivity, data aggregation and analysis, tracking and alerts, and remote medical assistance to detect and control infectious disease outbreaks. More researchers are now developing automated and effective alert systems for early and timely detection of outbreaks of such diseases to reduce morbidity and mortality and prevent global spread. These timely and effective public health measures are the essential requirements to avoid the risk of continuing outbreaks and the possibility of a local outbreak turning into a global pandemic such as COVID-19.

Safe Blues: A Method for Estimation and Control of the Spread in the Fight Against COVID-19 [29]

As is evident from the experience to manage the crisis, there is an immediate need for timely and accurate information about the spread of the COVID-19 virus. Such information can be used as input to the predictive models that could be used by different governments to help in their decision-making process. In general, the spread of a virus depends on both the biological properties of the virus and also on the behavioral properties of the populace.

Researchers have started doing research to study the biological properties of COVID-19 since the start of the outbreak. On the other hand, the enforced social distancing measures forces changes in the population behavior, which is very hard to observe and predict. As a result, providing a real-time estimate of the expected number of individuals infected by an infected person is very difficult. The error in estimation at different stages of the pandemic can result in an improper decision, affecting human life and having significant economic consequences.

"Safe Blues" team [29] assumes that in an imaginary world, one may think of creating a benign biological virus that has similar spreading properties as COVID-19 and is traceable through cheap and reliable diagnosis. Then by intentionally spreading such an imaginary virus among the population, the spread of COVID-19 could be easily estimated as the benign virus would respond to population behavior in a similar manner to COVID-19. However, such a benign biological virus does not exist, so the team propose a safe and privacy-preserving digital alternative to the benign and call it as "Safe Blues."

The Safe Blues method uses Bluetooth signals in a similar way of existing contact tracing apps. The existing contact tracing apps, deployed on personal mobile devices, are monitoring viral threats and collecting data about potential contacts. Such mobile apps can be adapted to transmit Safe Blues signals in a privacy-preserving manner.

The Safe Blues idea is that mobile devices mimic virus spread via the safe exchange of Bluetooth signals. The aggregated counts are reported to a server without any private information. The infection count can be estimated by periodically creating various strands of Safe Blues and repeatedly spreading them through the population using their smartphones. This estimation will help government and other organizations to plan their operations and be prepared. The estimated result is supposed to be a near-real-time measure depending on the level of social distancing.

At the moment, when government tries to impose different social distancing directives, it is not clear what the population compliance is? Even if compliance is being followed immediately, it often takes weeks to see the effect on the spread of COVID-19. In such a scenario, it is difficult for the government to decide an optimal social distancing measure. However, the Safe Blues idea will produce a much faster and precise feedback.

Government implementing "Safe Blues" during the COVID-19 pandemic obtains information during that period, which will be highly beneficial for decision making in the battle against COVID-19 and to mitigate second or third waves of attack. The experimental simulation analysis shows that Safe Blues data is able to predict COVID-19 infection within the asymptomatic population. For this, statistical machine learning and artificial intelligence methods, mixed with solid epidemiological models, are used to build estimates of COVID-19 spread as a function of live measurements of Safe Blues spread. It is expected that COVID-19 progresses, the estimates will become more and more precise.

At the end, government will be in a better position to know how to optimally "flatten the curve" or keep the curve below the healthcare system threshold. The implementation of Safe Blues within an existing contact tracing software platform is straightforward and does not require complicated software design. The "Safe Blues" model does not seek to develop an independent mobile app, but rather integrates the idea in the existing contact tracing applications.

Advisory for the Use of Sensors and IoT Framework

The advancements in the electronics, MEMS technology, sensor, wireless communication, and overall IoT framework are being utilized to address the issues and challenges arising out of the COVID-19 pandemic. Internet medical clinics and Webchat, extensive information investigations for contact tracing, use of BLE signal strength or GPS tags, distributed computing, Internet of things, artificial intelligence, robots, 5G telemedicine, and clinical data frameworks are extensively used to encourage clinical administration and management of COVID-19 patients and infections. Many countries, including China, USA, Singapore, etc., showed that wellbeing data innovations assume a significant job in reacting to the COVID-19 infection spread with the utilization of the IoT and sensors together with cloud computing infrastructure and artificial intelligence integrated machine learning [30–32].

This pandemic has set off an extraordinary interest in advanced wellbeing innovation arrangements. It has uncovered adequate mechanisms, such as for populace screening, following the disease, contact tracing, organizing the utilization of designated medical equipment, and planning focused on reactions.

Healthcare providers should take the travel history of all patients with respiratory indications and any international travel in the previous two weeks. They should set up an arrangement for the patients with the respiratory ailment in the outpatient department and give them necessary protective gear. The healthcare providers also should utilize protective covers themselves while analyzing such patients and practice hand cleanliness. Suspected cases ought to be shifted to government-assigned habitats for separation and testing.

Patients suffering from extreme pneumonia and deep respiratory trouble need to be assessed for travel history and put under contact and bead detachment. Regular purification of surfaces should be done immediately. They need to be tested for COVID-19 infection using different testing methods like the R-PCR test. Medical practitioners and administrators need to be vigilant and keep themselves aware of the advances in protocols, technology, and spread patterns around the world. Individuals and groups should practice the utmost responsibility and avoid spreading misinformation and rumors to mislead the general population.

References

1. Lin, B., Wu, S.: COVID-19 (Coronavirus Disease 2019): opportunities and challenges for digital health and the Internet of medical things in China
2. Joyia, G.J., Liaqat, R.M., Farooq, A., Rehman, S.: Internet of medical things (IOMT): applications, benefits, and future challenges in healthcare domain. J. Commun. **12**(4), 240–247 (2017)
3. Singh, R.P., Javaid, M., Haleem, A., Suman, R.: Internet of things (IoT) applications to fight against COVID-19 pandemic. Diabetes Metab. Syndr. Clin. Res. Rev. **14**(4), 521–524 (2020)
4. Haleem, A,, Javaid, M,, Vaishya, R,, Deshmukh, SG.: Areas of academic research with the impact of COVID-19. AJEM (Am. J. Emerg. Med.) (2020). https://doi.org/10.1016/j.ajem.2020.04.022
5. Swayamsiddha, S., Mohanty, C.: Application of cognitive Internet of medical things for COVID-19 pandemic. Diabetes Metab. Syndr. Clin. Res. Rev. (2020). ISSN 1871-4021
6. Vaishya, R., Javaid, M., Khan, I.H., Haleem, A.: Artificial intelligence (AI) applications for COVID-19 pandemic. Diabetes Metab. Syndr. Clin. Res. Rev. (2020)
7. Javaid, M., Vaishya, R., Bahl, S., Suman, R., Vaish, A.: Industry 4.0 technologies and their applications in fighting COVID-19 pandemic. Diabetes Metab. Syndr. Clin. Res. Rev. (2020). https://doi.org/10.1016/j.dsx.2020.04.032
8. Allam, Z., Jones, D.S.: On the coronavirus (COVID-19) outbreak and the smart city network: universal data sharing standards coupled with artificial intelligence (AI) to benefit urban health monitoring and management. In: Healthcare, vol. 8, no. 1, p. 46. Multidisciplinary Digital Publishing Institute (2020)
9. Yang, T., Gentile, M., Shen, C.F., Cheng, C.M.: Combining point-of-care diagnostics and Internet of medical things (IoMT) to combat the COVID-19 pandemic. Diagnostics (2020)

10. Wang, D., Kaplan, L., Le, H., Abdelzaher, T.: On truth discovery in social sensing: a maximum likelihood estimation approach. In: Proceedings of the ACM/IEEE 11th International Conferences on Information Processing in Sensor Networks (IPSN), pp. 233–244 (2012)
11. Wang, D., Amin, M.T., Li, S., Abdelzaher, T., Kaplan, L., Gu, S., Pan, C., Liu, H., Aggarwal, C.C., Ganti, R.: Using humans as sensors: an estimation-theoretic perspective. In: IPSN-14 Proceedings of the 13th International Symposium on Information Processing in Sensor Networks, pp. 35–46. IEEE (2014)
12. Zhang, D., Wang, D., Vance, N., Zhang, Y., Mike, S.: On scalable and robust truth discovery in big data social media sensing applications. IEEE Trans. Big Data (2018)
13. Zhang, D.Y., Wang, D., Zhang, Y.: Constraint-aware dynamic truth discovery in big data social media sensing. In: 2017 IEEE International Conference on Big Data (Big Data), pp. 57–66. IEEE (2017)
14. Yin, X., Han, J., Philip, S.Y.: Truth discovery with multiple conflicting information providers on the web. IEEE Trans. Knowl. Data Eng. 20(6), 796–808 (2008)
15. Chu, X., Ilyas, I.F., Krishnan, S., Wang, J.: Data cleaning: overview and emerging challenges. In: Proceedings of the 2016 International Conference on Management of Data, pp. 2201–2206 (2016)
16. Zhang, Y., Zong, R., Han, J., Zhang, D., Rashid, T., Wang, D.: Transres: a deep transfer learning approach to migratable image super-resolution in remote urban sensing. In: International Conference on Sensing, Communication, and Networking (SECON), p. to appear. IEEE (2020)
17. Shang, L., Zhang, D.Y., Wang, M., Lai, S., Wang, D.: Towards reliable online clickbait video detection: a content-agnostic approach. Knowl.-Based Syst. 182, 104851 (2019)
18. Li, H., Ota, K., Dong, M.: Learning IoT in edge: deep learning for the Internet of things with edge computing. IEEE Netw. 32(1), 96–101 (2018)
19. Zhang, D., Vance, N., Zhang, Y., Rashid, M.T., Wang, D.: Edgebatch: towards AI empowered optimal task batching in intelligent edge systems. In: 2019 IEEE Real-Time Systems Symposium (RTSS), pp. 366–379 (2019)
20. Zhang, D., Rashid, T., Li, X., Vance, N., Wang, D.: Heteroedge: taming the heterogeneity of edge computing system in social sensing. In: Proceedings of the International Conference on Internet of Things Design and Implementation, pp. 37–48 (2019)
21. Vance, N., Zhang, D.Y., Zhang, Y., Wang, D.: Privacy-aware edge computing in social sensing applications using ring signatures. In: 2018 IEEE 24th International Conference on Parallel and Distributed Systems (ICPADS), pp. 755–762. IEEE (2018)
22. Zhang, Y., Lu, Y., Zhang, D., Shang, L., Wang, D.: Risksens: A multi-view learning approach to identifying risky traffic locations in intelligent transportation systems using social and remote sensing. In: 2018 IEEE International Conference on Big Data (Big Data), pp. 1544–1553. IEEE (2018)
23. Zhang, Y., Wang, H., Zhang, D., Lu, Y., Wang, D.: Riskcast: social sensing based traffic risk forecasting via inductive multi-view learning. In: Proceedings of the 2019 IEEE/ACM International Conference on Advances in Social Networks Analysis and Mining, pp. 154–157 (2019)
24. Zhang, Y., Dong, X., Zhang, D., Wang, D.: A syntax-based learning approach to geolocating abnormal traffic events using social sensing. In: Proceedings of the 2019 IEEE/ACM International Conference on Advances in Social Networks Analysis and Mining, pp. 663–670 (2019)
25. Wang, D., Zhang, D., Zhang, Y., Rashid, M.T., Shang, L., Wei, N.: Social edge intelligence: Integrating human and artificial intelligence at the edge. In: 2019 IEEE First International Conference on Cognitive Machine Intelligence (CogMI), pp. 194–201. IEEE (2019)
26. Kim, Y., Huang, J., Emery, S.: Garbage in, garbage out: data collection, quality assessment and reporting standards for social media data use in health research, infodemiology and digital disease detection. J. Med. Internet Res. 18(2), e41 (2016)
27. Misinformation will undermine coronavirus responses. URL https://dailybrief.oxan.com/Ana lysis/DB250989/Misinformation-will-undermine-coronavirus-responses

28. Suman, R., Javaid, M., Haleem, A., Vaishya, R., Bahl, S., Nandan, D.: Sustainability of coronavirus on different surfaces. J. Clin. Exp. Hepatol. (2020)

29. Dandekar, R.J., Henderson, S.G., Jansen, M., Moka, S., Nazarathy, Y., Rackauckas, C., Taylor, P.G., Vuorinen, A.: Safe Blues: a method for estimation and control in the fight against COVID-19. medRxiv and bioRxiv, Apr (2020). https://doi.org/10.1101/2020.05.04.20090258

30. Lenert, L., McSwain, B.Y.: Balancing health privacy, health information exchange and research in the context of the COVID-19 pandemic. J. Am. Med. Inform. Assoc. (2020)

31. Ienca, M., Vayena, E.: On the responsible use of digital data to tackle the COVID-19 pandemic. Nat. Med. **26**(4), 463–464 (2020)

32. Reeves, J.J., Hollandsworth, H.M., Torriani, F.J., Taplitz, R., Abeles, S., Tai-Seale, M., et al.: Rapid response to COVID-19: health informatics support for outbreak management in an academic health system. J. Am. Med. Inform. Assoc. (2020)

Chapter 2
IoT and Sensor Network

Introduction

The IoT is a very much characterized plan between associated protested with sensors, advanced, and mechanical gadgets having the ability to transmit information over the remote medium without any or least social inclusion at any level. Every one of these gadgets is related to their novel distinguishing proof numbers or codes. IoT is presently settled and dependable innovation, which goes about as an intersection to the umpteen functionalities, quick examination, use of AI, and tangible items. IoT, alongside different advancements like cloud and AI, is of extraordinary use during the emergency.

In the current pandemic circumstance, the quantity of tainted patients is expanding universally—an earnest need to use the advances of the Internet of things technology and technique [1]. Furthermore, IoT has already been employed to serve the purposes in different domains such as the Internet of Healthcare Things (IoHT) or the Internet of Medical Things (IoMT), which are closely associated with the present pandemic [1].

IoT has four advance building blocks that are generally organized in a method (see Fig. 2.1). All of the four stages are related such that data is gotten or dealt with at one stage and yields the impetus to the accompanying stage. Facilitated characteristics in the process bring impulses and pass on robust business prospects [2].

Stage 1: The initial step includes sending interconnected devices that join sensors, actuators, screens, locators, and camera structures. These devices assemble data [2].

Stage 2: Usually, data obtained from sensors and various devices are in straight-forward structure, which ought to be amassed and changed over to the propelled structure for extra data dealing with [2].

Stage 3: Once the data is digitized and amassed, this is pre-arranged, standardized, and moved to the server or cloud [2].

S. K. Udgata and N. K. Suryadevara, *Internet of Things and Sensor Network for COVID-19*, SpringerBriefs in Computational Intelligence, https://doi.org/10.1007/978-981-15-7654-6_2

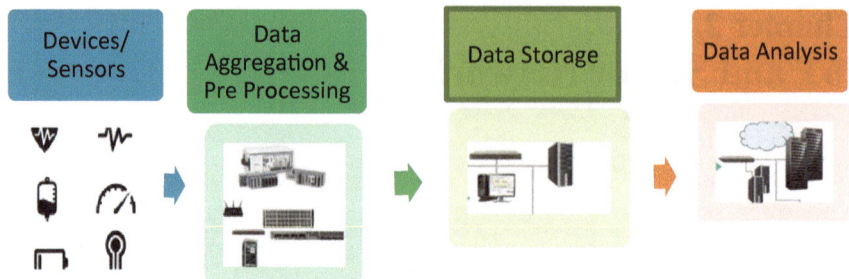

Fig. 2.1 IoT for healthcare applications

Stage 4: Final data is directed and dismembered at the essential level. Advanced analytics, applied to this data, brings significant business bits of information for a reasonable dynamic.

IoT is rethinking healthcare management by guaranteeing better consideration, improved treatment results, and decreased expenses for patients, and better procedures and work processes, improved execution, and patient experience for human services suppliers [2].

IoT is an innovative technology that can be used to identify infected persons due to this virus, quarantine, and monitor them during the quarantine. All high-risk patients can be tracked easily using the Internet-based network [1]. This technology can also be used for biometric measurements like blood pressure, heartbeat, and glucose level [1]. Some future applications envisioned for IoT sound like science fiction, be that as it may, a bit of the more suitable and down to earth sounding open doors for the advancements incorporate:

- receiving alarms on the phone or wearable contraption when IoT frameworks perceive some physical danger is recognized close by,
- self-parking vehicles,
- an automatic mentioning of food supplies and other home supplies,
- automatic following of action affinities and other ordinary individual activity including the target keeping and standard headway,
- Smart home and environmental monitoring conditions [3, 4].

Layers of IoT Architecture

The "Things" in IoT terminology, for the most part, alludes to IoT gadgets/devices that have stimulating characteristics and can accomplish distant monitoring, inducing, and read-through abilities.

IoT gadgets/devices can [5]:

- Exchange data with other related contraptions/gadgets and applications (honestly or by suggestion).

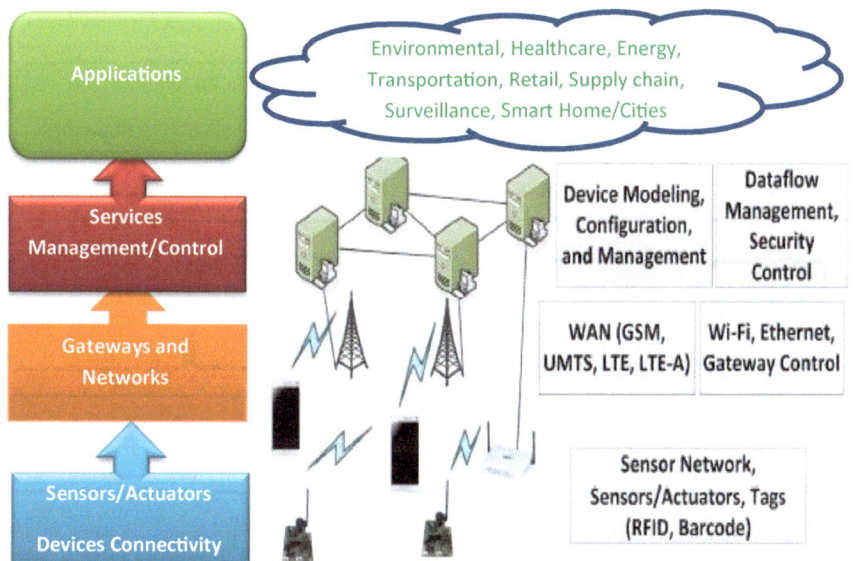

Fig. 2.2 Generic IoT architecture layers

- Collect data from various devices/gadgets and process the data either locally or send the data to unite servers or cloud-based application back-closes for taking care of the data.
- Perform a couple of tasks locally and various endeavors inside the IoT structure, considering objectives.

There are four significant layers in the underlying IoT architecture, as shown in Fig. 2.2.

In general, the IoT architecture, the base layer, consists of the sensors and their connectivity arrangement, which collects data from the environment/entities. At the next higher point, we have the gateway and network layer which coordinates with the devices for effective communication. The next segment is the management service layer, and afterward, toward the top, we have the application layer where the information gathered is handled by the requirements of different applications.

The basic features of each layer:

Sensors Connectivity

This layer involves RFID devices, actuators, and sensors (a central bit of an IoT system). This structures the principal "things" of an IoT system.

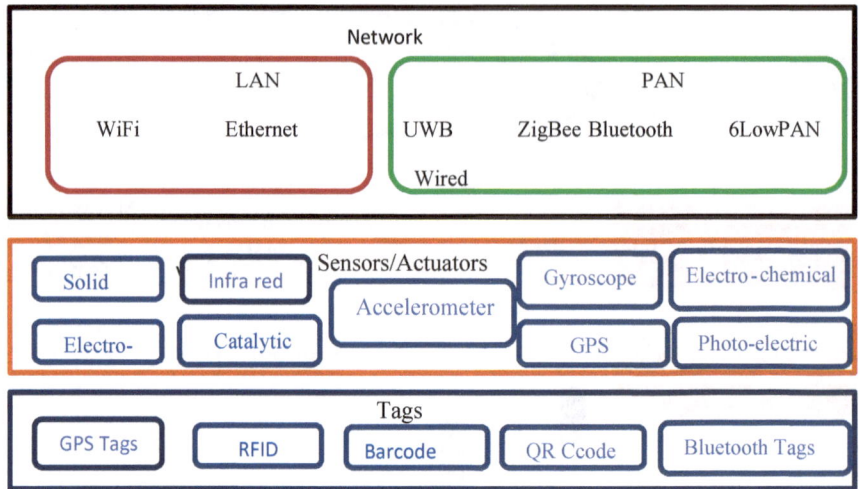

Fig. 2.3 Sensors and connectivity layer

- Sensors and RFID marks are remote devices that structure the wireless sensor networks (WSN).
- Sensors are dynamic, which suggests that constant information is to be accumulated and processed.
- This layer in like manner has the framework accessibility (like WAN, PAN, etc.), liable for passing on the rough data to the accompanying layer, which is the gateway, and network layer.
- The contraptions which contained WSN have restricted capacity, less transmission limit, and have small taking care of speed.
- We have different sensors for different applications—temperature sensor for ambient environment temperature data, water quality for taking a look at water quality, and humidity sensor for atmosphere humidity or soil, etc.

As per Fig. 2.3 underneath, at the base of this layer, we have the names which are the RFID devices, sensors/actuators, and a while later the correspondence frameworks.

Gateway/Network Layer

This layer guides the data starting from the sensor, and their connectivity is passed to the accompanying layer, the management service layer.

- This layer requires having a capacity limit concerning taking care of the colossal proportion of data accumulated by the sensors, RFID marks, etc. Moreover,

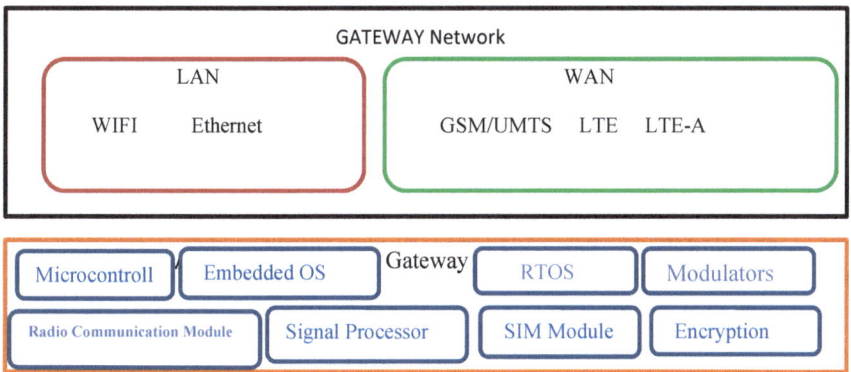

Fig. 2.4 Gateway/Network layer

this layer needs to have dependably trusted in execution concerning various frameworks.

- Different IoT contraption goes after different kinds of framework shows. These shows are required to be caught up in an appropriate system. This layer is liable for fusing distinctive framework shows.

From Fig. 2.4 underneath, at the base, we have the portal, which is incorporated embedded OS, signal processors, modulators, microcontrollers, etc. Over the entry, we have the gateway networks which are local area network (LAN), wide area network (WAN), etc.

Management Service Layer

This layer is utilized for dealing with the IoT administrations. The service layer is responsible for securing analysis of IoT gadgets, analysis of information (stream analytics, data analytics), and device management.

- To extricate the fundamental data from the gigantic measure of crude information gathered by the sensor gadgets to yield an important after effect of the considerable amount of information, this activity is performed in this layer [6].
- This layer also helps in doing that by abstracting information, removing unwanted data, and dealing with the information stream.
- This layer is additionally answerable for information mining, content mining, data/network administration, and so on.

The administration layer has an operational support service (OSS), which incorporates device modeling, device configuration, and management. Likewise, from the figure, we can see that there are IoT/M2M application services, which incorporates data analytics. Security incorporates access controls, encryption, identity access management, and so on; and afterward, we have the business rule management (BRM) and business process management (BPM).

Application Layer

The application layer shapes the highest layer of IoT engineering, which is answerable for compelling usage of the information gathered.

Different IoT applications incorporate home automation, E-wellbeing, E-government, and so forth.

IoT Protocols

The protocols that are used in the IoT theme can be related to the TCP/IP model as shown below in Fig. 2.5.

Link-Layer Protocols [7]

Link-layer conventions fundamentally decide how the information is sent over the system's physical layer or medium. The hosts on a similar connection will trade the information utilizing these conventions. It additionally decides how the data transporters are coded and motioned by the equipment gadget over the medium to which the host is appended. A portion of the link-layer conventions are as per the following: IEEE standards 802.3 Ethernet: 802.3, which speaks to an assortment of wired Ethernet norms for the connection layer [7].

Examples:

- 802.3—Ethernet (10 Mbps–40 Gbps, coaxial, twisted pair)
- 802.11—WiFi (1 Mbps–6,75 Gbps, wireless)
- 2G/3G/4G—Mobile communications
- 802.16—WiMax (1.5 Mbps–1 Gbps, wireless)
- 802.15.1—Bluetooth (wireless)
- 802.15.4—LR-WPAN (40–250 Kbps, wireless).

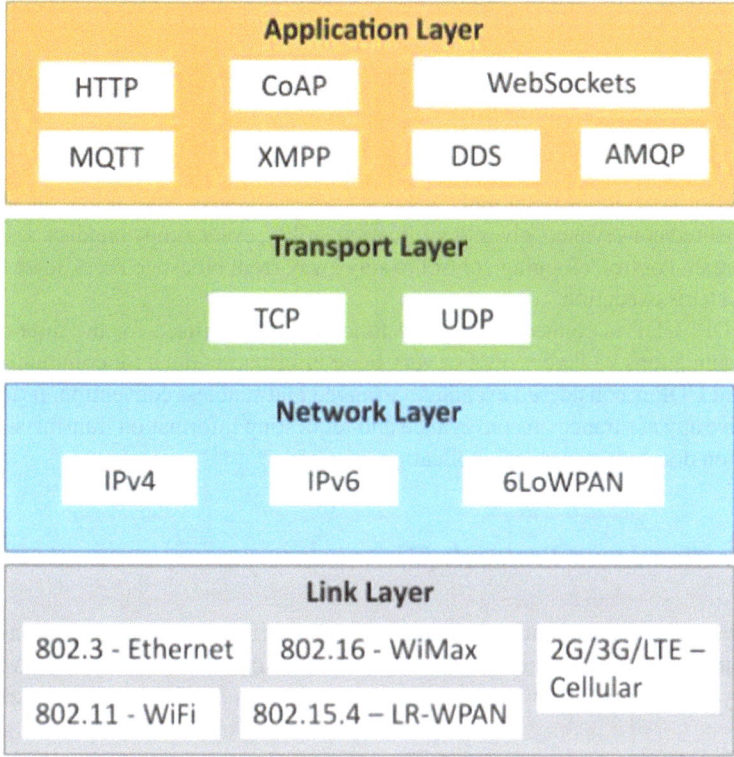

Fig. 2.5 IoT protocols stack

Network/Internet Layer Protocols [7]

This layer is liable for transmitting the data or the information from source to goal, by playing out the host tending to and bundle directing. The recognizable host proof should be possible by progressive IP plans, for example, IPv4 or IPV6.

(1) IPV4: Most conveying Internet convention, which is primarily utilized for recognizing gadgets in the system utilizing a 32-bit address, permits 4,294,967,296 gadgets/devices to be identified uniquely. The reliability of data transmission is not guaranteed.

(2) IPV6: Successor of IPV4 utilizes 128-bit address permits to assign 3.4 * 1038 devices/gadgets.

(3) 6LowPAN: IPV6 low-power personal area networks (6LowPAN), which carries IP conventions to the lower power gadgets that have restricted handling capacity, works on ISM 2.4 GHz frequency and gives information paces of 250 Kb/s.

Transport Layer Protocols [7]

These vehicle layer conventions are answerable for the total conveyance of message exchanges between devices/gadgets. The ability of massage moves can be set up on associations utilizing TCP by handshaking or without affirmation as in UDP.

(1) TCP: TCP is an essentially a connection arranged convention; guarantees ensured conveyance, gives identification ability, evades copy bundles, and gives stream control, clogging control to stay away from blockage breakdown during systems execution.

(2) UDP: UDP is connectionless and fundamentally utilized for the time-critical applications with very little or very huge information units for communication, and UDP is considered exchange-arranged and stateless convention. It does not give any assurance on conveyance and requesting information transmission and even does not ensure on duplication.

Application Layer Protocols [7]

Application layer conventions fundamentally characterize how the applications interface with the lower layer conventions to send information over the system. Application layer conventions empower procedures to process associations by utilizing the ports.

(1) HTTP: Hypertext transfer protocol is a stateless convention, and the convention fundamentally follows a solicitation reaction model; every HTTP demand is autonomous of different solicitations. HTTP utilizes general asset identifiers to recognize HTTP assets.

(2) COAP: Constrained application protocol is intended for M2M (machine) applications. The CoAP is an online transmission convention for utilization with obliged hubs and compelled organizes in IoT. CoAP runs on connectionless-UDP as opposed to association situated TCP, which utilizes client--server design for transmission.

(3) WebSocket: Web socket communication utilizes full-duplex correspondence over single attachment association with exchange of the messages among customers and servers. It fundamentally utilizes TCP, the surge of messages is sent among back and forth among customers and servers, and the association is kept open. Customers can be an IoT gadget, program, or any portable application.

(4) MQTT: Message queue telemetry transport transfers on distributing buy-in model, and it is a lightweight informing convention. It utilizes customer server design where the customer is an IOT gadget interface with the server (MQTT dealer) and afterward distributes the messages to themes on the server. The representative advances the messages to customers bought into points. The intermediary advances the messages to the customers by buying into the themes. MQTT

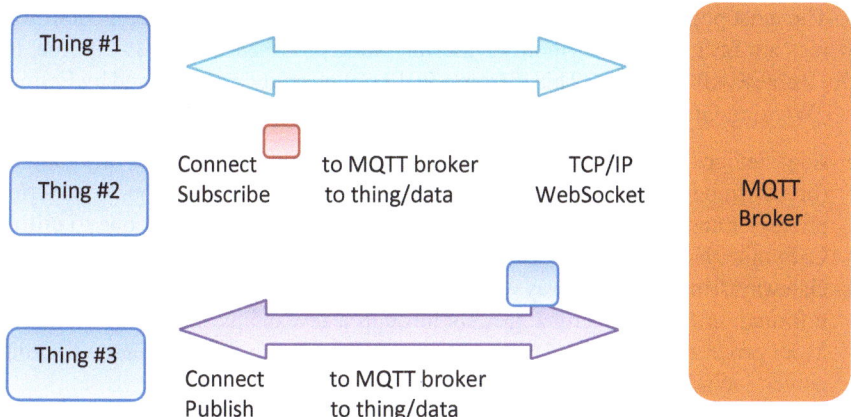

Fig. 2.6 MQTT-publish/subscribe model

is reasonable for restricted handling gadgets and when the system transmission capacity is low. The underlying MQTT operational mechanism is shown in Fig. 2.6.

(5) XMPP: Extensible messaging and presence protocol (XMPP) is utilized for correspondence in real-time and pushing XML information between various gadgets/devices. XMPP covers a broad scope of utilizations like information syndication, informing, gaming, multi-party visit, and voice/video calls. XMPP is considered as a decentralized convention which bolsters two-path correspondences among customer and server. XMPP permits correspondence between various IoT gadgets [7].

(6) DDS: Data distribution service (DDS) is a middleware standard considered as information-driven, which is mostly utilized for the gadget-to-gadget correspondence or a machine-to-machine correspondence. DDS fundamentally utilizes distribute supporters model; the points were made by distributors for which endorsers can buy-in. The distributer is fundamentally an article which is liable for the circulation of information, and supporter is liable for accepting distributed information. DDS gives the nature of administration control and configurable dependability.

IoT Programming Languages and Coding [8]

Programming for IoT usually is a combination of different programming languages since the Internet of things (IoT) is an arrangement of between related gadgets that are used with extraordinary features and the capacity to move information over a multi-system. The decision of programming language relies upon the capacity and reason for the gadget/device used in the IoT theme. IoT envelops an assortment of gadgets, including edge devices/gadgets, gateways/routers, and cloud servers.

The most popular programming languages considered in the development of IoT themes are Java, C, C++, Python, Javascript, Node.js, Assembler, PHP, C#, Lua, R, Go, Ruby, Swift, and Rust. This is from a 2017 online study co-supported by Eclipse IoT Working Group, IEEE IoT, AGILE IoT, and IoT Council.

- Edge devices—These are compelled asset installed frameworks. For tiny gadgets (higher imperatives), Assembly and C are the dialects of decision. A superior processor and all the more figuring power on the gadget empower one to utilize C, Python, Node.js; Java is used to boost execution speed.
- Gateways/Routers—Gateways oversee correspondence and do an examination of information from numerous gadgets through a few unique modes of transport. More programming can be run on these gadgets due to their expanded figuring power, including C, C++, Java, Python, and Node.js.
- Cloud—With the about boundless registering capacity that is accessible, structures like Apache, Hadoop, and HiveQL can figure and process massive IoT datasets. Factual registering and representation should be possible utilizing programming languages like R or Julia.

Hub-centric programming approach—Every angle is modified by the IoT engineer, like correspondence between hubs, assortment, and examination of sensor information, giving orders to actuator hubs. Communication between gadgets is expressly encoded.

Database approach—Every hub is viewed as a significant aspect of a database, and the designer can give inquiries to sensor hubs. The concentrate is on assortment, conglomeration, and sending of information to the base station.

Functional programming—Abstractions are given to determine significant level correspondence. The IoT gadgets data will be collected and applied various functions to infer the meaning from the sensor data.

Model-driven turn of events—The sensor fusion of the IoT data will be applied to the model (AI) to reason the meaning generated from the sensor data.

The features of IoT programming language [8]:

Versatility—Programming systems that help assorted programming designs can perform load-adjusting progressively.

Simultaneousness—Real-time correspondence between a large number of gadgets and applications means a great many simultaneous associations.

Coordination—Programming language supports for unequivocally (control is driven) or verifiably (information-driven) arranging the job of processing components.

Heterogeneity—Programming structure gives direction on how calculations are mapped to the figuring components.

Adaptation to internal failure—Applications ought to have the option to nimbly go from online to disconnected state as systems parcel and mend their associations.

Light impression—Regarding runtime, overhead, and as far as programming exertion, the structure ought to be light.

Backing for inactivity and affectability—In geologically conveyed applications, pushing all calculations to the cloud is not perfect. The programming system needs to deal with these necessities powerfully.

Over the previous decade, the prevalence of Python as a standard programming language has detonated. Eminent points of interest of Python over different dialects incorporate, yet are not restricted to;

1. It is a straightforward language to learn and simple to execute and convey so does not have to invest a great deal of energy, adapting bunches of organizing principles and arranging choices.
2. It is compact, expandable, and embeddable, and in this way, it is not framework subordinate and thus bolsters a great deal of single-board PCs available nowadays, regardless of engineering and working framework.
3. Most significantly, it has a large network that gives a ton of help and libraries for the language.

Python packages for IoT: The Python bundles utilized for creating IoT applications.

- mraa:

mraa is a skeleton GPIO library for most SBCs which bolster Python. The beneficial thing about it is that there is only one library for all gadgets, so there is no need to utilize different ones for an Edison and a Pi. An elevated level library, perusing from and writing to pins, is a one-line issue, and the library likewise offers help for correspondence conventions, such as I2C, UART, and SPI.

- sockets

A socket is a bundle which encourages organizing over TCP/IP and UDP utilizing Python. It gives access to Berkeley attachment APIs to get to the Internet. Both TCP/IP and UDP are transport layer conventions, and they are perfect for correspondence with gadgets on a similar WiFi arrangement. One of the additionally fascinating employments of attachments, in my experience, is that one can manufacture their correspondence convention utilizing this bundle as the base.

- mysqldb

A database is an easy decision concerning most IoT applications. For something whose sole reason for existing is to send information to the Web, there should be a database, at any rate, a remote one that stores this produced information. MySQL is the go-to social database for most engineers. In such a manner, mysqldb is an exceptionally advantageous little device that goes around the need to execute shell orders inside a Python content to peruse and keep in touch with a database.

- NumPy

Having utilized MATLAB broadly during undergrad examines, I have become used to managing exhibits. Python, then again, manages records as a substitute for the cluster, which is equivalent to having a birch tree supplant your rottweiler as the watchman of the house. It only does not work. Fortunately, NumPy is there to get you out. It is, generally, a bundle for logical figuring utilizing Python, fundamentally the same as MATLAB, yet a lot lighter. The element I utilized most is to peruse sensor information in mass from my databases and work on them utilizing the built-in capacities.

- matplotlib

Information representation is one of the most significant activities that can be performed. It looks entirely unusual when you convert a colossal rundown of numbers to a short diagram which can be seen instinctively. It is additionally beneficial if you happen to be an academician. You know how significant those diagrams can be in distribution. matplotlib gives various styles of diagrams that can be plotted utilizing nearby information.

- pandas

Another library for information researchers, pandas, is a bundle devoted to information investigation. It is basically a neighborhood option in contrast to utilizing SQL databases, which are increasingly fit to managing information as it is based on NumPy. It has numerous favorable circumstances over the previous, for example, an increasingly smoothed out way to deal with information taking care of and examination, direct process on nearby datasets, and the capacity to deal with heterogeneous and unordered information.

- OpenCV

The older sibling of sign handling, picture preparing, was customarily the area of superior, exclusively fabricated equipment. Despite everything, such gadgets do the activity a lot quicker than their single-board partners; it is in any event a chance. Furthermore, in circumstances where versatility and availability are organized over speed, this may simply be the answer for those uncommon occasions. OpenCV is a Python port of the fruitful C library for picture preparation. It contains significant level variations of natural picture preparing capacities, which make photograph examination a lot simpler.

- tkinter

Even though this library comes preinstalled with all establishments of Python, it still deserves notice. Tkinter is a GUI improvement library that comes packaged in with all appropriations of Python. For individuals who are more OK with a cut injury instead of item arranged programming, figuring out how to utilize this bundle might be somewhat overwhelming from the start; however, the prizes more than compensate for the exertion. Each part of your Python content can be controlled, utilizing an impromptu GUI. This is amazingly valuable in the circumstances, such as user testing or rehashed execution of a similar code.

- TensorFlow

TensorFlow is a bundle for numerical calculations for AI. It uses an alternate scientific portrayal called information stream charts, which use hubs as numerical activities and edges as information exhibits. This is a precious library to have if you manage a ton of non-direct datasets or work broadly with choice trees and neural systems.

- requests

HTTP is one of the significant conventions utilized in customary Web-based asset trade, being progressively fit toward large information trades. The solicitations bundle is utilized in Python to make HTTP calls and parse reactions. This bundle is valuable when managing HTTP-based outsider cloud administrations.

- paho-MQTT

MQTT is a convention grown exclusively for the Internet of things worldview. Its emphasis on rapid correspondence for low payload correspondence between asset compelled gadgets. The paho-MQTT library gives a straightforward understanding of the adaptation of the convention for use with implanted frameworks. MQTT solicitations can be made legitimately inside Python, with no special arrangement to be finished, particularly valuable in the prototyping stage.

IoT Implementation and General Usage: [17–20]

The major components of IoT implementation are:

(I)   Sensors/Electronic devices
(II)   Networks (Wired/Wireless communication)
(III) Standards
(IV)   Smart environment
(V)   Smart analysis of data
(VI)   Smart actions/responses
(I)   Electronic gadgets/devices.

As indicated by (IEEE), sensors can be characterized as: an electronic gadget that produces electrical, optical, or computerized information got from a state of being or occasion. Information created from sensors is then electronically changed, by another gadget, into data (yield) that is helpful in dynamic done by "astute" gadgets or people (individuals).

Kinds of sensors: active sensors and passive sensors. The determination of sensors significantly affected by numerous components, including:

reason (temperature, motion, bio…, and so on.), precision, unwavering quality, range, goals, and level of intelligence

The main thrusts for utilizing sensors in IoT today are new patterns in innovation that make sensors less expensive, more astute, intelligent, and smaller in size. Challenges for smart sensing systems are less energy consumption, robustness, security, and interoperability

(II)   Networks (Wired/Wireless communication)

The subsequent stage is transmitting the information assembled by sensors over frameworks with all the different fragments of a run, including switches and interfaces in different geographies, including LAN, MAN, and WAN. Interfacing the different bits of frameworks to the sensors should be conceivable by different advances including WiFi, Bluetooth, low-power WiFi, Wi-Max, standard Ethernet, long-term evolution (LTE), 5G, 6G, and the progressing promising development of Li-Fi (using light as an instrument of correspondence between the different gadgets/devices).

(III)   Widely acknowledged arrangement of IoT conventions and principles [9–16]

MQTT: Message queue telemetry transport (MQTT) is a lightweight show for sending direct data streams from sensors to applications and middleware. The show limits on TCP/IP and fuses three sections: endorser, wholesaler, and vendor. The distributer accumulates data and sends it to endorsers. The mediator tests distributers and endorsers, checking their endorsement and ensuring security. MQTT suits pretty much nothing, humble, low-memory, and low-power contraptions.

DDS: Data distribution service is an IoT standard for steady, adaptable, and world-class machine-to-machine correspondence. It was made by the object management group (OMG). You can pass on DDS both in low-impression devices and in the cloud. The DDS standard has two essential layers: information-centric publish-subscribe (DCPS), which passes on the information to endorsers, and information local reconstruction layer (DLRL), which gives an interface to DCPS functionalities.

AMQP: Advanced message queuing protocol is an application layer shown for message-masterminded middleware conditions. It is embraced as an overall standard. The getting ready chain of the show fuses three sections that cling to explicit rules.

Bluetooth: Bluetooth is a short-run interchanges innovation incorporated into most phones and cell phones, which is a significant preferred position for individual items, especially wearables.

Bluetooth is quite useful for versatile clients. In any case, in the not so distant past, the new noteworthy convention for IoT applications showed up Bluetooth Low Energy (BLE) or Bluetooth Smart. This innovation is an excellent establishment for the IoT, as it is adaptable and adaptable to all market advancements. Also, it is intended to lessen power utilization.

Standard: Bluetooth 4.2

Recurrence: 2.4 GHz

Range: 50–150 m (Smart/BLE)

Information rates: 1Mbps (Smart/BLE)

Zigbee: 3.0 is a low-power, low information rate remote system utilized generally in modern settings. The Zigbee alliance even made the universal language for the Internet of things—Dotdot—which makes it feasible for brilliant items to work safely on any system and flawlessly see one another.

Standard: ZigBee 3.0 dependent on IEEE802.15.4

Recurrence: 2.4 GHz

Range: 10–100 m

Information rates: 250 kbps

WiFi: WiFi is the innovation for radio remote systems administration of gadgets. It offers quick information move and can process much information. This is the most popular sort of availability in LAN situations.

Standard: Based on IEEE 802.11

Frequencies: 2.4 and 5 GHz groups

Range: Approximately 50 m

Information rates: 150–200 Mbps, 600 Mbps greatest

Cell: Cell innovation is the premise of cell phone systems. However, it is likewise reasonable for the IoT applications that need working over longer separations. They can exploit cell correspondence capacities, for example, GSM, 3G, 4G, and 5G. The innovation can move high amounts of information, yet the force utilization and the costs are high as well. Hence, it very well may be an ideal answer for ventures that send modest quantities of data.

Standard: GSM/GPRS/EDGE (2G), UMTS/HSPA (3G), LTE (4G)

Frequencies: 900/1800/1900/2100 MHz

Range: 35 km (GSM); 200 km (HSPA)

Information rates: 35–170 kps (GPRS), 120–384 kbps (EDGE), 384 Kbps–2 Mbps (UMTS), 600 kbps–10 Mbps (HSPA), 3–10 Mbps (LTE)

LoRaWAN: Long-range wide area network is a convention for vast region systems. It is intended to help large systems (for example, shrewd urban communities) with a massive number of low-power gadgets. LoRaWAN can give ease portable and secure bidirectional correspondence in different businesses.

Standard: LoRaWAN

Recurrence: Various

Range: 2–5 km (urban region), 15 km (rural territory)

Information rates: 0.3–50 kbps

The Internet of things has become the premise of computerized change and robotization, growing new business contributions, and improving how we live, work, and engage ourselves.

(IV)   Smart environment monitoring:

Weather monitoring: IoT-based climate checking frameworks can gather information from various sensors appended (for example, temperature, mugginess, pressure, and so forth.) and send the information to cloud-based applications and capacity back-closes.

The information gathered in the cloud can then be broke down and pictured by cloud-based applications. Climate cautions can be sent to the registered clients from such applications.

AirPi is a climate and air quality observing unit equipped for recording and transferring data about temperature, mugginess, pneumatic force, light levels, UV levels, carbon monoxide, nitrogen dioxide, and smoke level to the Internet.

Air pollution monitoring: Partly based air contamination checking frameworks can screen outflow of destructive gases (CO_2, CO, NO, NO', and so on.) by manufacturing plants and cars utilizing vaporous and meteorological sensors. The gathered information can be broke down to settle on educated choices on contaminations control draws near.

Noise pollution monitoring: Because of developing urban turn of events, commotion levels in urban areas have expanded and even gotten alarmingly high in some urban regions. Commotion contamination can cause wellbeing dangers for people because of rest disturbance and stress. Commotion contamination checking can help in creating clamor maps for urban communities. Urban commotion maps can help the arrangement producers in urban arranging and making strategies to control clamor levels close to local locations, schools, and stops.

Parcel-based clamor contamination observing frameworks utilize various commotion checking stations that are created at better places in a city. The information on clamor levels from the stations is gathered on servers or in the cloud. The gathered information is then totaled to create clamor maps.

Forest fire detection: Woods flames can make harm typical assets, property, and human life. There can be various reasons for backwood fires, including helping, human carelessness, volcanic ejections, and sparkles from rock falls. Early discovery of backwoods flames can help in limiting the harm.

Parcel-based backwoods fire recognition frameworks utilize various observing hubs conveyed at various areas in timberland. Each checking hub gathers estimations on surrounding conditions, including temperature, moistness, light levels, and so on.

River floods detection: Stream floods can make broad harm healthy and human life. Waterway floods happen because of constant precipitation, which causes the stream levels to increase and stream rates to increment quickly. Early alerts of floods can be given by observing the water level and stream rate.

Parcel-based stream flood observing framework utilizes various sensor hubs that screen the water level (utilizing ultrasonic sensors) and stream rate (utilizing the stream speed sensors).

Information from various such sensor hubs is totaled in a server or in the cloud. Checking applications raise alarms when fast increment in water level and stream rate is distinguished.

(V)   Smart analysis of data [15, 16]

The sensor data in the IoT theme are transmitted, fused, processed, and analyzed. The following are instances of the sorts of information IoT gadgets gather:

Robotization information. Numerous individuals are suspicious of gadget robotization. Regardless of whether it is computerized lights in an office or programmed settings on an indoor regulator, robotization is vital. Without robotization, somebody's activity would make sure to modify the indoor regulator settings two times per day, and the last one out turns off all the lights.

Status information. The most essential and common kind of IoT information is status information. Most IoT gadgets produce status information, which is gathered as crude information, and afterward utilized for increasingly complex examination.

Area information. Consider area information as an indoor worldwide situating framework: Rather than guiding you to a particular goal, area information empowers you to follow bundles and gear progressively.

Handling IoT Data

The enormous measure of information that IoT sensors and gadgets create must be handled before the data can be utilized. Nonetheless, because the information regularly originates from various gadgets or in various configurations, there are a few things you should do before handling or applying any kind of examination to the collected information:

- Standardize or change the information to a uniform organization, guaranteeing that arrangement is good with your application.
- Store or make a reinforcement of the recently changed organization.
- Filter any unwanted, obsolete, or undesirable information to help improve precision.
- Integrate extra organized (or unstructured) information from different sources to help advance your present informational collection.

IoT Data Analytics [15, 16]

IoT examination is performed by applying information investigation apparatuses or systems to the different sorts of information IoT gadgets create. Utilizing IoT investigation, unimportant data can be removed from enormous information assortments that would then be able to be utilized to enhance systems, applications, business procedures, and creation. A few sorts of information examination can be utilized on IoT information:

Prescriptive investigation. The prescriptive investigation is utilized to break down steps to take for a particular circumstance. It is frequently portrayed just like a blend of illustrative and prescient investigation. At the point when utilized in business applications, prescriptive investigation disentangles many data to acquire progressively exact ends.

Spatial examination. This technique is utilized to dissect area-based IoT information and applications. Spatial examination interprets different geographic examples, deciding any sort of spatial connection between different physical items. Leaving applications, keen vehicles, and yield arranging are generally instances of utilizations that profit by the spatial investigation.

Streaming examination. Spilling examination, now and then alluded to as occasion stream preparing, encourages the investigation of huge "moving" informational collections. These constant information streams can be investigated to distinguish crisis or earnest circumstances, encouraging a quick reaction. The sorts of IoT information that profit by spilling investigation incorporate those utilized in rush hour gridlock examination, fair dealing, and the following of budgetary exchanges.

Time arrangement investigation. Time arrangement investigation depends on time-sensitive information, and information is breaking down to uncover any peculiarities, examples, or patterns. Two frameworks that enormously advantage from time arrangement investigation are wellbeing checking and climate observing frameworks.

References

1. Singh, R.P., Javaid, M., Haleem, A., Suman, R.: Internet of things (IoT) applications to fight against COVID-19 pandemic. Diabetes Metab. Syndr. Clin. Res. Rev. (2020)
2. https://www.wipro.com/en-IN/business-process/what-can-iot-do-for-healthcare-/
3. Mukhopadhyay S.C., Suryadevara N.K.: Internet of things: challenges and opportunities. In: Mukhopadhyay, S. (ed.) Internet of Things. Smart Sensors, Measurement, and Instrumentation, vol. 9. Springer, Cham
4. Suryadevara, N.K., Mukhopadhyay, S.C.: Smart Homes: design, implementation, and issues. Springer, Cham, Switzerland (2015)
5. Ray, P.P.: A survey on Internet of things architectures, J. King Saud Univ. Comput. Inform. Sci. (2018)
6. Jadhav, D., Chobe, S.V., M. Vaibhav, M., Khandare, L: Missing person detection system in IoT. In: 2017 International Conference on Computing, Communication, Control and Automation (ICCUBEA), 2017
7. Niraja, K.S., Murugan, Prabhu, C.S.R.: Security risks in internet of things: a survey. In: IEEE International Conference on Computational Intelligence and Computing, Research (ICCIC), 2017
8. Sethi, P., Sarangi, S.R.: Internet of things: architectures, protocols, and applications. https://doi.org/10.1155/2017/9324035
9. Farahzadi, A., Shams, P., Rezazadeh, J., Farahbakhsh, R.: Middleware technologies for cloud of things—a survey. Digit. Commun. Netw. Amsterdam
10. Lashkari, B., Rezazadeh, J., Farahbakhsh, R., Sandrasegaran, K.: Crowdsourcing and sensing for indoor localization in IoT: a review. IEEE Sens. J. 19(7), 2408–2434 (2019)
11. Srisooksai, T., et al.: Practical data compression in wireless sensor networks: a survey. J. Netw. Comput. Appl. 35(1), 37–59 (2012)
12. Rezazadeh, J., Subramanian, R., Sandrasegaran, K., Kong, X., Moradi, M., Khodamoradi, F.: Novel iBeacon placement for indoor positioning in IoT. IEEE Sens. J. 18(24), 10240–10247
13. Čolaković, A., Hadžialić, M.: Internet of things (IoT): a review of enabling technologies, challenges, and open research issues. Comput. Netw. 144, 17–39 (2018)
14. Lee, I., Lee, K.: The Internet of things (IoT): applications, investments, and challenges for enterprises. Bus. Horiz. 58(4), 431–440 (2015)
15. Kelly, S.D.T., Suryadevara, N.K., Mukhopadhyay, S.C.: Towards the implementation of IoT for environmental condition monitoring in homes. IEEE Sens. J. 13(10), 3846–3853 (2013)

16. Suryadevara, N.K., Mukhopadhyay, S.C.: Determining wellness through an ambient assisted living environment. IEEE Intell. Syst. **29**(3), 30–37
17. https://www.iotforall.com/glossary-iot-standards-and-protocols/
18. https://standards.ieee.org/initiatives/iot/stds.html
19. https://www.etsi.org/technologies/internet-of-things
20. Agarwal, P., Alam, M.: Investigating IoT middleware platforms for smart application development. =https://arxiv.org/abs/1810.12292

Chapter 3
COVID-19, Sensors, and Internet of Medical Things (IoMT)

Introduction

The IoT empowered automatons are being used for observation to guarantee user isolation and face mask usage to cover the face adequately. These innovations can be utilized for following and identifying the source of an outbreak. It can be useful to the disease transmission specialists to identify persistent zeroeth patients and to recognize the people interacting with these patients. The patients who break the isolation can likewise be found. Additionally, this innovation can be a potential candidate for giving help to the clinical staff by remote monitoring and observing in-home patients and staying away from physical contact.

The IoT is utilized for different applications to satisfy the significant necessity of countering the COVID-19 pandemic impacts. It can evaluate the upcoming circumstance with the assistance of acquired information. Its uses are applied for legitimate administration of this pandemic. The patients can utilize the IoT framework for reliably observing the pulse rate, circulatory strain, glucose level, and different exercises for customized consideration. It serves the purpose of the ceaseless screening of the wellbeing states of more individuals. One of the critical uses of this innovation is to follow the constant area of clinical gear and devices for a smooth and effective treatment process. This makes the treatment work process of the patient effective and supports dynamic procedures for complex cases. The IoT paired with artificial intelligence (AI) technology can be deployed to automate or scale tasks, including patient tracking, quarantining, monitoring, and in-hospital care, as shown in Fig. 3.1.

A self-governing IoT empowered police robot can be used for watching the regions to affirm that the individuals are following the lockdown conventions appropriately. The self-governing police robots can be sent to the medical clinics to help the wellbeing laborers play out their obligations with no interruption—this aide in improving the presentation of the clinical staff and containing the spread of the COVID-19.

S. K. Udgata and N. K. Suryadevara, *Internet of Things and Sensor Network for COVID-19*, SpringerBriefs in Computational Intelligence, https://doi.org/10.1007/978-981-15-7654-6_3

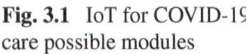
Fig. 3.1 IoT for COVID-19
care possible modules

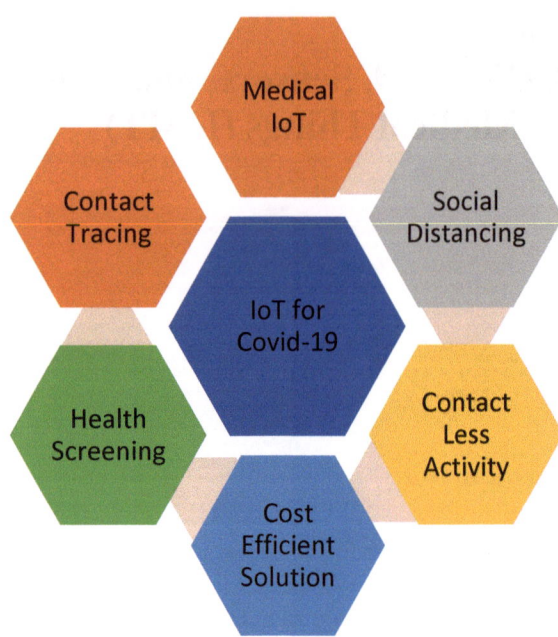

Sensor Systems for COVID-19

Regarding COVID-19, relevant information about patient healthcare includes information on doctor--patient arrangements, clinical pictures, doctor notes, case history, chest X-ray reports, and data about episode zones, among few others. This information is produced from various sources, starting from the Internet of things (IoT) sensors (e.g., smartphone information) to online social stages (e.g., open responses). The conventional information systematic apparatuses and components are not satisfactory for meeting the prerequisites during the COVID-19 pandemic. The sensor system for medical services/applications can have the following generic stages, as shown in Fig. 3.2.

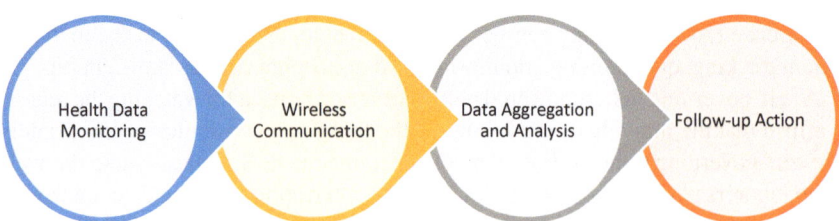

Fig. 3.2 IoT-based sensor system for health monitoring and action architecture

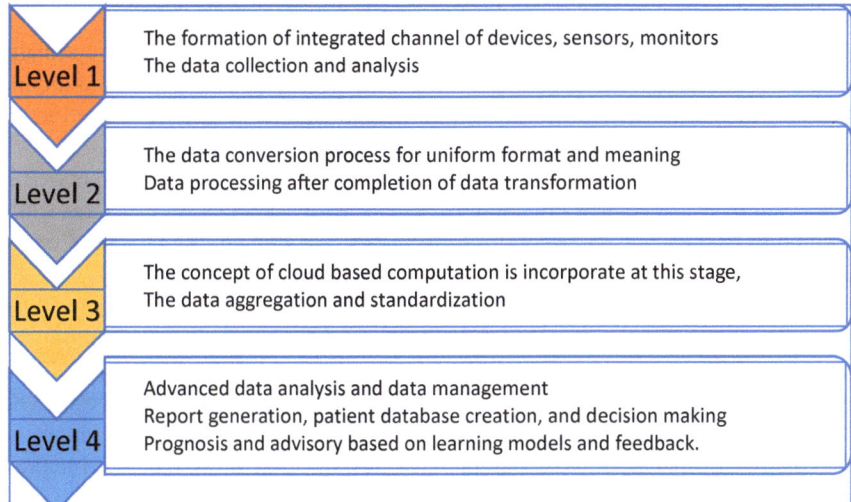

Fig. 3.3 Multi-level process of IoT sensor data processing

Modern smartphones are embedded with a large number of sensors and possess powerful computing facilities. It is possible to sense information about daily activities and also to capture visual data using smartphones. Smartphones are capable of capturing, collecting, communicating, and storing large volume data from either suspected and confirmed COVID-19 patients. A smartphone can scan CT images of a COVID-19 patient and upload them to the cloud for analysis purposes. Even multiple CT images of the same COVID-19 patient can be examined by the smartphone for comparative analysis to determine how lesions have developed over a period of time. The investigation is beneficial to the suspected COVID-19 cases to diagnose and monitor the grade of lung inflammation. A multi-level sensor data processing model is presented in Fig. 3.3.

Specialists at the Swiss Federal Laboratory have built up a sensor that might rapidly and dependably distinguish SARS-CoV-2, i.e., the COVID-19 infection. Preceding the coronavirus spreading far and wide, Jing Wang and his group investigated sensors that could detect microscopic organisms and infections that were noticeable. The idea of the sensor technology would not supplement the setup research facility tests, yet could be utilized as an alternate strategy for a clinical analysis to quantify infection in the air in a continuous manner. The sensor could be used in crowded and sensitive places like train stations and emergency clinics.

The sensor highlights optical and thermal impacts to identify the infection reliably and securely. It depends on small structures of gold nanoparticles on a glass substrate and DNA receptors that coordinate explicit RNA groupings of SARS-CoV-2. The receptors on the sensor are similar arrangements to the coronavirus' interesting RNA successions to help reliably distinguish the infection. Wang and the group of scientists utilized restricted surface plasmon reverberation to detect the infection. The innovation uses an optical marvel that occurs in metallic nanostructures. They optimally balance the incident frequency that reaches to make a close plasmonic field around the nanostructure. When particles bind to the surface, the neighborhood refractive index changes in the close plasmonic field and indicates the infection [1].

Nations have demonstrated that it is possible to lessen the spreading of the SARS-CoV-2 or novel coronavirus infection by executing robust defensive measures while attempting to prop the economy up. In the USA, the Centers for Disease Control and Prevention (CDC) suggests the day-by-day wellbeing checks of employees before they enter the worksite. The World Health Organization (WHO) likewise recommends the temperature screening at the working environment. One of these measures is the utilization of non-contact infrared thermometry temperature estimation to screen for raised skin temperatures (EST) in the working environment. Employees and guests can have their temperature screened before they enter the workplace. This is one of the apparatuses to keep manufacturing plants, structures, building locales, organizations, or offices of any sort to open for regular business and ensure that everybody stays safe.

Infrared sensors and thermal cameras screen for elevated skin temperature. Skin temperature is not the same as the body temperature but can be accepted as the body temperature under normal environmental conditions [2].

Disease Tracking Through Crowdsensing

Crowdsensing-based disease tracking (CDT) requires sensor networks and groups of people or volunteers (crowd) with mobile/smartphone devices capable of sensing, collectively sharing disease-related information (e.g., early symptoms, nearby infected persons, deciding to self-quarantine). The success of CDT is mainly dependent on the observation that individuals tend to proactively volunteer in contributing data about the COVID-19 spread using their smartphones, wearables, or other devices with sensors and available communication connectivity. CDT is relatively less pervasive and requires the active participation of people and physical hardware sensors in contrast to the Social-media-driven Disease Spread Indicator (SDSI [3]) [4]. However, the data in the CDT framework is less noisy and hence more reliable. Figure 3.4 [5] shows an example of a representative CDT system. A CDT may typically incorporate three main components.

1. Data collection platform: Data collection framework is the first component which consists of a connected network of users with a customized smartphone application to log observed data, and a set of IoT devices (e.g., activity tracker, smart heart rate monitors, thermal scanners). The smartphone applications allow authorized users to contribute their reports about COVID-19 actively. If the users choose to input data, the mobile app lets them conjure at what granularity (e.g., state, county, street, house) they are comfortable sharing their location information.

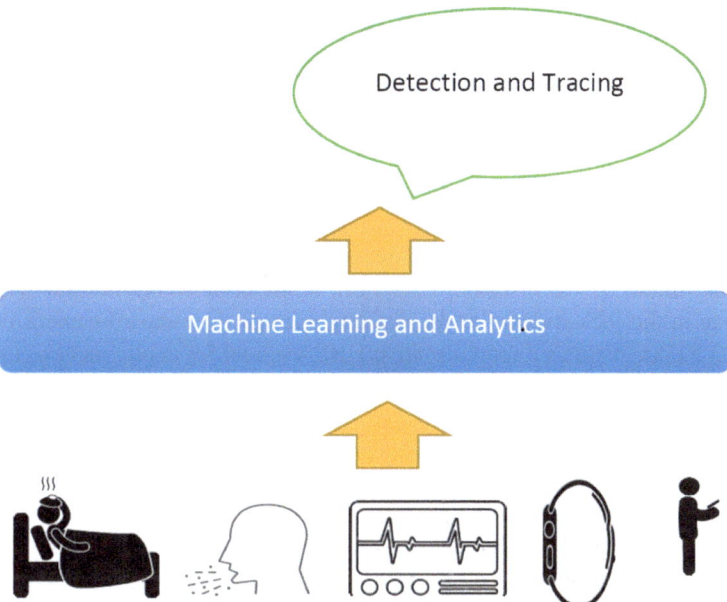

Fig. 3.4 Overview of the CDT system

2. Data analysis framework: The analytics framework applies relevant statistical analysis and artificial intelligence (AI) techniques together with machine learning (ML) on the collected data to infer probable regions of infection and relatively safe zones. To conserve bandwidth and faster expedite processing, the smartphones' computational power can also be harnessed to execute the AI algorithms at the edge.
3. Smartphone application framework: The smartphone application framework on the end-users' mobile phones visually represents the analyzed geospatial distribution of the inferred regions. The app can obtain the required information from the backend server depending on the user queries (e.g., checking the risk level of a particular area of interest). In most cases, the data collection, processing, and representation are carried out in the same smartphone application. The Singapore and South Korea governments have launched mobile apps that utilize crowd-sourced data to trace community transmission of the COVID-19. The government of India also launched a similar application (although not precisely the same) named "Arohya Setu" for crowdsourced contact tracing.

UAV-Based Health Surveillance and Alerting

Unmanned ariel vehicle (UAV)-based health surveillance and alerting (UHSA) systems have emerged as a new dimension of solutions to mitigate the challenges that evolved due to the urgency of the COVID-19 outbreak. With the help of many onboard sensors (e.g., optical and thermal cameras, GPS, accelerometer, gyroscope, microphones), UAVs can gather information remotely during a disease pandemic scenario where the ground units and human patrol teams cannot operate due to risks of getting infected. For instance, UAVs can assist in detecting unwanted crowds of people in the locked-down areas of a city through the onboard camera images. Figure 3.5 demonstrates a representative UHSA model for deployment to mitigate the COVID-19 spread [4].

The UHSA system responds to individuals' emergency requests through social media posts or user-operated dedicated IoT devices about unnecessary mass gatherings. The data obtained from the UAVs and users in a backend server is processed using social sensing approaches based on statistical analysis, deep learning, and machine learning for analyzing and ascertaining the truthfulness of the data. The information is then updated across nearby regions by sending messages, raising verbal alerts, or alarm sirens through speakers installed on the UAVs. UAVs are also dispatched to different areas of a city to scan and obtain situational information about the region proactively. Using the onboard sensors and image classification algorithms like convolutional neural networks (CNNs), UHSA detects whether people violate the rules during the lockdown situation (e.g., by roaming outside, gathering in crowds, not following prescribed social distancing). The framework may also locate

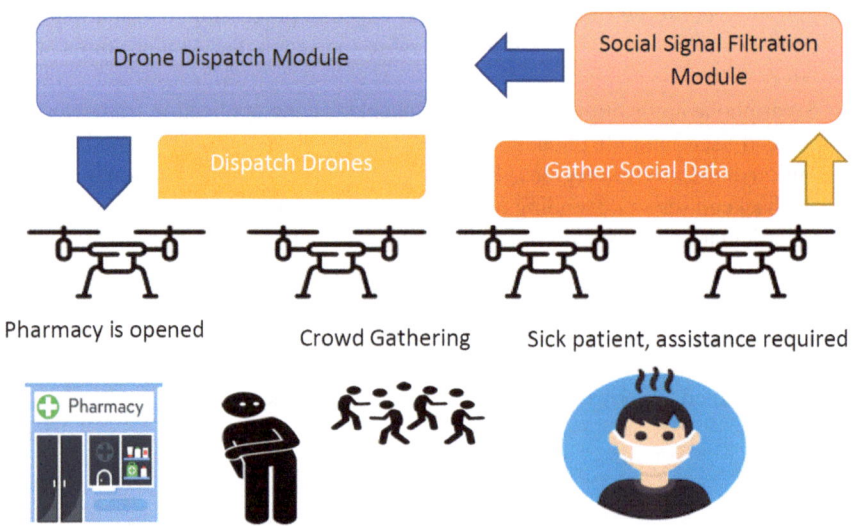

Fig. 3.5 UAV-based health surveillance and alerting system

and verify the availability of critical supplies using the UAVs (e.g., open pharmacy, grocery stores) based on social media posts. Using the onboard speakers of the UAVs, the people violating the rules are advised to return home.

A real-world example of UHSA during the COVID-19 ordeal is in California, USA, where the law enforcement officials have resorted to utilizing drones for patrolling the state of California during the ongoing lockdown situation. During the COVID-19 crisis in China, UAVs have served multiple roles, including post-epidemic aerial evaluation, alerting, and relief distribution to affected regions. The representative model of the UHSA system is shown in Fig. 3.5.

Ongoing IoT Research for Addressing COVID-19 Challenges

1. Empowering and enforcing social distancing measures utilizing smart city and the associated IT infrastructures:

A COVID-19 pandemic has uncovered and exposed the limitations of the current organizations. Accordingly, engineering, technology applications, and innovation frameworks should be leveraged for a quick and effective response to the COVID-19 and similar challenges. This section attempts to describe the novel engineering, potential use cases, and some possible future advances in developing such applications utilizing the smart city and its available infrastructure [3]. COVID-19 outbreak is unprecedented and has disrupted the lives of millions of people across the world and paralyzed the economy. This pandemic has opened several research challenges and opportunities that our community is capable of and must address to equip itself to mitigate the present crisis and the future [3]. The smart city IT infrastructure includes edge devices, a reliable communication module, storage and computation cloud services (both private and public), intelligent computing platform, informed and smart users. There are many devices like the CCTV network, widespread availability of smartphones, smartwatches, various type of GPS-enabled tracking devices, Bluetooth-enabled devices available in the smart city IT infrastructure. All these devices collect massive data that can be analyzed through video processing, proximity detection, people counting to find out whether social distancing, face cover is being strictly followed or not?

2. Internet of Medical Things (IoMT) for remote monitoring of patients

The IoMT is a special case of the Internet of things specially designed for medical applications and services to assist individuals and medical practitioners. IoMT can help in screening, detecting, informing and advising the patients, caregivers, health administrators, and doctors to provide proper healthcare and ensure timely and genuine information dissemination and contain the issues before they become serious. In this sense, the authors [5] proposed a model, namely Assisted Reproduction Treatment (ART), to decrease the number of medical clinic visits, diminish medicinal services costs, improve patient care, and reduce the cost and inconvenience. Thus, an IoMT-based innovative proposition to oversee and control the solution of pharmacological medicines to patients who have completed and used ART formats [5].

One of the significant benefits of IoMT-based remote wellbeing monitoring is the possibility of providing remote assisted living and wellbeing monitoring of the patients under continuous observation in isolation, combined with the advantage of low medical clinic bills. The ordinary remote monitoring frameworks cause lots of inconvenience and discomfort to the patients because of the size and form factor of the modules to be wearable in the body and the frequent charging or replacement of batteries. The IoMT concept's evolution addresses these issues by designing and developing compact, ultra-low power sensor modules, low power and small-sized controllers, and lightweight communication protocols.

The remote wellbeing checking system mainly consists of a dynamic and flexible patient observing unit (PPMU) at the patient's home or emergency clinical assistance vehicles and ongoing checking with the support of available networks at the hospital. The PPMU mainly consists of sensors and electronic circuits which are good and suitable for acquiring vital parameters. The vital parameters are pulse rate, pulse changeability, breath rate, systolic circulatory strain, diastolic blood pressure, oxygen immersion, internal heat level, body mass index, total lung volume, blood glucose level, and a few other parameters. It also has a processing unit to process the received sensor outputs and a communication submodule to uploads to the server for additional investigation and analysis.

The schematic diagram of PPMU in the patient's home or an emergency vehicle is shown in Fig. 3.6 [6]. Doctors can always use a good and useful user-friendly

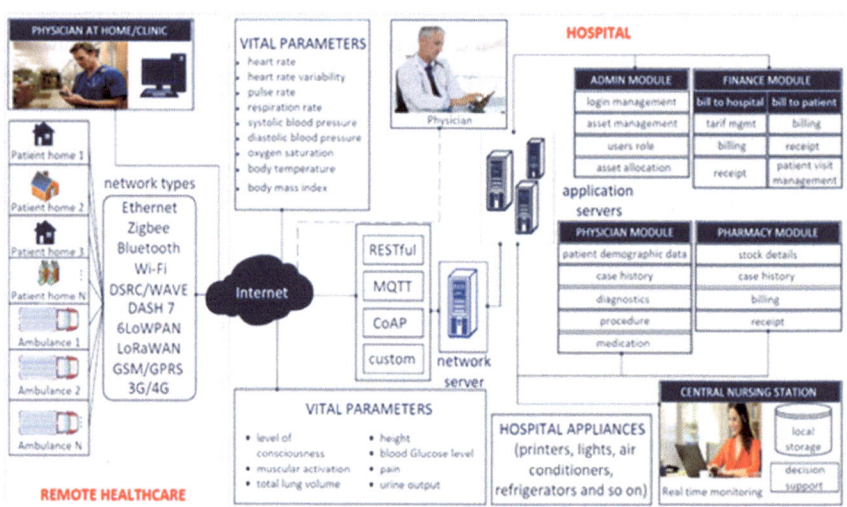

Fig. 3.6 Generic architecture of IoMT [6]

graphical user interface (GUI) to monitor and check patients' state. The general remote checking frameworks depend on the type of disease and associated infection pattern [6].

"When to eat, what to eat, how much to eat, and how often to eat" are the most important questions one ponders for ascertaining good health. Not able to decide on these in an optimum manner leads to several medical problems. In iLog [7, 8], a model is proposed to screen and inform the client about both the quantity and quality of the food. iLog collects data on the medical condition of an individual alongside the eating practices. Based on this information, it can classify normal-eating or stress-eating concerning every individual. The proposed iLog model is a learning model that can naturally recognize, order, and evaluate the food items from the plate of the client using a few sensors and camera images. It uses different state-of-the-art AI and machine learning models together with user feedback for refining the learning process. The iLog model can perform with a maximum accuracy of 98% with an average accuracy of 85.8% [8].

As user conditions and requirements keep changing with respect to time, the IoMT framework is also required to be flexible, dynamic, and adaptive to perform satisfactorily. Authors in paper [9] proposed an adaptive IoMT framework and separately test for people who are relatively less mobile and also in a situation where users are permitted to move freely. This proposed model also acts as an IoMT-based catastrophe rescuer wellbeing observing framework, which works as a rescue manager with a search facility in case of a debacle [9]. This framework is illustrated in Fig. 3.7.

3. Smartphone-based methods for infection detection

There are different means to detect coronavirus that includes clinical investigation of chest CT examination pictures and blood test results. The COVID-19 patient shows various symptoms like fever, sluggishness, and dry hack. Keeping this in view, a few

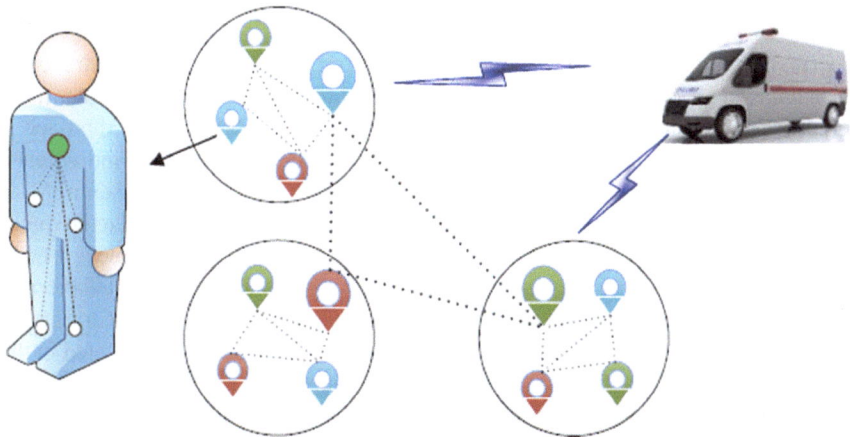

Fig. 3.7 IoMT architecture for health monitoring system [9]

strategies can be adopted to develop a clinical recognition kit to identify the infection. However, these dedicated kits will be of prohibitive cost and convenience of use and acceptability by the users. Authors in paper [10] proposed an alternative system to detect COVID-19 infection using the smartphone sensors. This alternate proposition rightfully assumes that a majority of users and caregivers have smartphones for day-to-day activities. The same smartphone can additionally be utilized for infection identification purposes also. These days Smartphones are equipped with existing computation rich processors, large enough memory spacé, and a large number of sensors, including cameras, amplifiers, temperature sensors, inertial sensors, proximity sensors, movement detection sensors, and stickiness sensors. Using the sensors of the smartphones, an artificial intelligence (AI) empowered system is developed for signal processing and estimations, and the evaluation of the seriousness of pneumonia [10].

The smartphone's reasonably accurate temperature sensor is used to measure the temperature of harmed tissues with its neighboring tissues to assess the difficult to recuperate wounds in patients with a diabetic foot ulcer (DFU) [11]. The change in heat conductivity estimated the tissues' properties and can be utilized for the early detection of DFU. This model makes use of the smartphone temperature sensor to get self-appraisal by patients and self-check diabetic conditions [11]. Medical authorities in Tehran (Iran) classified the patients into three groups using AI-based reasoning and machine learning. Group-1 contains patients with aggravation. Group-2 contains patients with vascular inconveniences, and Group-3 includes patients who were figuring out how to control their glucose levels to an adequate level, as per their clinical records and data collected from their smartphone periodically. Besides, the tissue conditions of indoor patients are also considered in the computational model. The temperatures of the ulcer and adjoining tissues are continuously measured and communicated to the central server. The gathered information is utilized in a proposed model for detecting infections of human tissues [11].

4.   IoT-enabled thermal screening for avoiding potential spread:

Thermal screening using thermal cameras or infrared thermometers are utilized at open areas to check the temperature level of individuals/groups to detect potentially infected persons and segregate them to avoid infection spreading. This approach is quite useful as it saves a lot of effort and time required to check each individual's temperature using traditional contact-based temperature measurement systems. Most importantly, the chance of the person in charge of screening getting infected reduces significantly as the screening is done in a contactless manner. This principle is used in the proposed framework [12]. It utilizes keen head protectors with a mounted thermal imaging system to identify and recognize potential COVID-19 infected persons from the thermal image without human intervention. The thermal camera is located in the keen cap with IoT architecture that can continuously perform the screening procedure and update the information in the central server. The proposed framework is also integrated with the facial acknowledgment module to recognize the infected person.

This model is quite successful and popular among healthcare providers, administrators, and managers to manage and stop spreading the infection to a larger extent [12].

COVID-19 has put extraordinary pressure on the world's health systems, exposes the populations' vulnerability, and critically threatens the global communities in an unprecedented manner. Although efforts are underway for detecting the virus, providing treatments and developing vaccines, it is also critically important to leverage available technologies and innovate newer technologies to control disease emergence, arrest its spread, and especially develop disease prevention strategies. The objective is to review enabling technologies and systems with various application scenarios for handling the COVID-19 crisis. This brief will focus specifically on

- Wearable devices suitable for monitoring the vulnerable populations at risk of infection in general and those in quarantine, both for evaluating the health status of caregivers and management personnel and for facilitating processes for admission into hospitals.
- Unobtrusive sensing systems for detecting the disease and for monitoring patients with mild symptoms, whose clinical situation may worsen, and
- Remote monitoring and diagnosis of COVID-19 infection and similar diseases using telehealth technologies [13].

The focus of research groups, institutions, and technological companies

According to Forrester research [14], healthcare providers will be a great compliment among IoT specialists once the interruptions brought about by the COVID-19 fade away. According to Achim Granzen, head investigator, Forrester research, just 7% of the work that significant IoT contribution is on smart healthcare [14]. Granzen also mentioned that this emergency has set off many thoughts and arrangements, and created an ideal opportunity to do more advanced studies on devices, methodology, and procedures. This will focus on governments, human services suppliers, healthcare providers, and administrators when the intense emergency is behind us. The work in this direction is already underway as the "drive-by" testing offices in South Korea, detecting and recognizing disease groups and contact tracing in Singapore.

Videoconferencing is playing an important role and currently pervasive, which assists with social separating while at the same time keeping organizations running and trying to do their usual business. We are experiencing a quicker adaptation in many different domains also. Schools are changing to video-classes, and even social occasions like marriages, family functions are broadcasted on the Web, and people prefer to participate remotely. Overall, the innovation is prepared and accessible to help fight this emergency and allowing work to be carried in a usual manner as far as practicable. Countries like Hong Kong and Singapore use this technology for travelers with a high probability of infection and monitoring those people on "stay home" notice.

Industry 4.0 drives abilities for remote tasks, checking and support of creation lines and assembling plants. This can have a fundamental influence on making producers in the Asia Pacific (APAC) stronger owing to the disturbances. A large portion of the measures has been set up rather impromptu, similar to a Proof of Concept (PoC). A good part of those PoCs is expected to turn into well-performing products for crisis management situations. Ensuring human lives is the foremost priority in all emergency reactions and preparedness models. Government and private organizations are striving hard to leverage the experience gained from this COVID-19 pandemic.

Effect of COVID-19 on IoT Activities

There are both positive and negative effects on the IoT activities due to the current COVID pandemic. With the infection outbreak and spread, there is also a danger that current IoT activities are affected as the manufacturing pants are closed, expertise is under lockdown, and operational destinations are shut down. For example, the delay of the 2020 Olympics Games is a blow for IoT suppliers who have developed new IoT-based models that are to be utilized during the games. Many efforts and investments have been made by Japanese and many IoT-based organizations in building cutting-edge IoT innovations for intelligent transport management systems, providing world-class hospitality, monitoring wellbeing and health, monitoring the proceedings of the game, and many other things. The delay in Olympics games denies those organizations an opportunity to exhibit on the world's most spectacular stage soon.

The world is also experiencing a surge in demand for newer use cases and applications in IoT domains during these unfortunate emergency circumstances like the COVID-19 emergency. Quickly, it adapted to the changing demands and responded well to the call by addressing the innovation bottlenecks, and limiting the computationally heavy and energy-hungry processes. IoT will continue to assume a significant role in modernizing human services and fiasco avoidance, open wellbeing, security, gracefully chain, assembling, and creation."

Due to the stringent social distancing measures to control the spread of infection, organizations will robotize their procedures more in the coming months. Before this emergency, the use of robots was viewed as a threat to our jobs and employment opportunities. The current pandemic has changed the horizon, and the use of robots is seen as a prudent alternative to human-intensive tasks. Experts and business innovators are looking for answers to the questions like "in what manner would automation be able to quicken our recuperation and shield us from future pandemics?". Robots working close by people in clinics to perform cleaning and mechanized distribution centers have very well demonstrated how people can be secured through computerization. Because of the severe effect of the emergency on flexible chains, the COVID-19 IoT effect prompts organizations to change their graceful chain procedures.

Numerous providers and organizations will reassess their single-provider and the single-nation policy that helps them reduce expenses when things are steady. The pandemic shows that it is excessively dangerous on occasions such as these to depend on one or barely any sources.

Healthcare is obviously at the center of the COVID-19 pandemic, and it is expected to experience a massive surge in demand. IoMT is attracting a lot of attention, and researchers from different domains, including electronics, communication, computer science, and mechatronics, are coming together to deliver the best. AI-enabled edge devices with low-cost, low-power communication technology integrated with cloud computing are revolutionizing the healthcare landscape.

Capital Expenses (CAPEX) to Operating Expenses (OPEX) will be a new model. In a Q4/2019 IoT Analytics study of makers, 58% showed that they would want to acquire hardware as help or rent instead of keeping it on their asset report. Amid an emergency, this transformation of CAPEX to OPEX makes it a lot simpler for organizations to downsize their expenses as request vanishes.

Many advancements and innovations are now flooded to address individual's and groups' wellbeing with the outbreak of COVID-19. These advancements mainly incorporate:

- Telehealth conferences. Telehealth (where a specialist treats the patient through a video gathering and offers guidance) has emerged as a viable alternative. Both healthcare provides and healthcare seekers feel secured. For instance, the Stanford Children's Health Hospital is currently experiencing many computerized visits each day.
- Computerized diagnostics. The natural next stage of telehealth is the use of sensors and IoT devices to perform computerized diagnostics. Clinics are exploring different avenues to explore this opportunity in this direction. Kinsa systems developed connected thermometers that are being used to control the spread in the USA.
- Remote checking. Remote checking and observing for providing assisted living to the senior citizens are experiencing a steep rise in demand. Livongo Health, a healthcare company, is focused on providing a line of products for remote IoT-based monitoring and management of "incessant sicknesses" to face the challenge arising from COVID-19 pandemic.
- Robot help. Many countries and especially in China, robots have been utilized to disinfect and clean emergency clinics and perform/remind medication schedules.
- New uses for drones: Drones and UAVs are there for quite some time and companies, and specialists have been working to innovate many use cases. This COVID-19 pandemic has suddenly opened up new use cases like mass screening, ensuring social distancing, and contact tracing. The following is a set of examples of the use of drones for mitigating pandemic challenges.

1. Clinical conveyances: According to Zhao Liang, COO of Antwork, during the previous month, the automaton conveyance framework in Xinchang County has helped nearby emergency clinics for distributing drugs and clinical reports with the help of more than 300 drones during this pandemic situation.
2. Surveillance and monitoring: Many countries have utilized drones and UAVs to screen open spaces and large gathering successfully.
3. Message and information dissemination: Drones have been used to broadcast messages and spread awareness among citizens. Drones use alarm sirens to

alert a group of users violating social distancing norms. Drones are also used to communicate information periodically.

4. Disinfecting and splashing: Horticultural automaton company XAG Co. Ltd. automated disinfectant sprayers and splashing using drones in the last few months.

Many users require to track the vessels for efficient supply chain management, and it seems to quite useful and effective. The lockdowns have caused significant disturbances, and vessel tracking has become more meaningful. The company vessel-tracker tries to provide technological solutions to provide a framework integrated with effective GUIs to track vessels and make sense if a vessel is going to their destination port, or if there has been a significant change to its expected time of arrival (ETA).

Suppliers of IoT innovation are utilizing their information and refreshing people in general on what is happening in their systems. Vesseltracker.com, for instance, as of late distributed updates (in German) on worldwide voyage boats and cargo movement. Geotab can generate reports on the business street transportation movement across North America.

South Korea, which can script success stories about dealing with COVID-19 challenges, utilized its "Savvy City Data Hub" to permit epidemiological specialists to demand, acquire, and affirm information about coronavirus cases and contact tracing.

References

1. https://www.medicaldesignandoutsourcing.com/this-sensor-could-detect-covid-19-in-the-air/
2. https://serverscheck.com/arrangements/crownCOVID-19.asp
3. Gupta, M., Abdelsalam, M., Mittal S.: Enabling and enforcing social distancing measures using smart city and its infrastructures: a COVID-19 Use case. arXiv preprint arXiv:2004.09246 (2020)
4. Rashid, M.T., Wang, D.: COVIDSens: a vision on reliable social sensing based risk alerting systems for COVID-19 spread. arXiv preprint arXiv:2004.04565 (2020)
5. Domínguez, D., Morales, L., Sanchez, N., Navarro-Pando, J.: IoMT-driven eHealth: a technological innovation proposal based on smart speakers. In: International Work-Conference on Bioinformatics and Biomedical Engineering. Cham: Springer
6. Vishnu, S., Ramson, S.J., Jegan, R.: Internet of medical things (IoMT)-An overview. In: 2020 5th International Conference on Devices, Circuits, and Systems (ICDCS). IEEE (2020)
7. Sayeed, M.A., Mohanty, S.P., Kougianos, E., Zaveri, H.: iDDS: an edge-device in IoMT for automatic seizure control using on-time drug delivery. In: 2020 IEEE International Conference on Consumer Electronics (ICCE), pp 1–6. IEEE (2020)
8. Rachakonda, L., Mohanty, S.P., Kougianos, E.: iLog: an intelligent device for automatic food intake monitoring and stress detection in the IoMT. IEEE Trans. Consum. Electron. (2020)
9. Wei, K., Zhang, L., Guo, Y., Jiang, X.: Health monitoring based on the internet of medical things: architecture, enabling technologies, and applications, pp. 27468–27478. IEEE Access (2020)
10. Maghdid, H.S., Ghafoor, K.Z., Sadiq, A.S., Curran, K., Rabie, K.: A novel AI-enabled framework to diagnose coronavirus COVID-19 using smartphone embedded sensors: design study. arXiv preprint arXiv:2003.07434 (2020)

11. Maddah, E., Beigzadeh, B.: Use of a smartphone thermometer to monitor thermal conductivity changes in diabetic foot ulcers: a pilot study. J. Wound Care **29**(1), 61–66 (2020)
12. Mohammed, M.N., Syamsudin, H, Al-Zubaidi, S., Rusyaizila Ramli, S.A.K, Yusuf, E.: Novel COVID-19 detection and diagnosis system using IoT based smart helmet. Int. J. Psychosoc. Rehabil. **24**(7), 2020
13. Ding, X.-R., Clifton, D., Nan, J.I., Lovell, N.H., Bonato, P., Chen, W., Yu, X., et al.: Wearable sensing and telehealth technology with potential applications in the coronavirus pandemic. IEEE Rev. Biomed. Eng. https://doi.org/10.1109/rbme.2020.2992838
14. https://futureiot.tech/iot-developers-to-focus-more-smart-healthcare-post-covid-19/
15. https://iot-analytics.com/the-impact-of-COVID-19-on-the-internet-of-things/

Chapter 4
Advances in Sensor Technology and IoT Framework to Mitigate COVID-19 Challenges

Introduction

The backbone of an IoT system consists of sensor networks that can recognize the hints of COVID-19. Upon identification, an area could be "secured" to restrict spread and guarantee brief treatment to contaminated people. It is not very difficult to envision such frameworks being fused into future smart city arrangements, which now incorporate applications planned for improving wellbeing with the help of an IoT theme.

IoT, along with the advancements in wireless communications, cloud technologies, and artificial intelligence (AI), has been in use for a comprehensive set of applications during the COVID-19 crisis. The low energy consumption, the network of short-distance communications technologies, and industry startups are re-engineered to design, develop, and prototype low-cost IoT smart sensing systems to assist, collaborate, and mitigate the global COVID-19 pandemic [1].

A few IoT and sensor-based systems in the forefront and are used to mitigate COVID-19 challenges are discussed in the following sections.

System for Avoiding Face Touch

Scientists at Media Lab, Massuehcest Institute of Technology (MIT), USA [2], have been developing a device to alert users against touching their faces. It is a seemingly simple device that works on the principle of wireless communication, magnetic flux detection, and machine learning integrated with artificial intelligence techniques. Since coronaviruses survive for days on many surfaces, a person can get COVID-19 by touching a contaminated surface or object and then touching their mouth, nose, or eyes. Face touching in a subconscious mind is quite natural among human beings, and quitting this habit is far easier said than done. Most people touch their face

© The Editor(s) (if applicable) and The Author(s), under exclusive license to Springer Nature Singapore Pte Ltd. 2021
S. K. Udgata and N. K. Suryadevara, *Internet of Things and Sensor Network for COVID-19*, SpringerBriefs in Computational Intelligence,
https://doi.org/10.1007/978-981-15-7654-6_4

frequently throughout the day, usually without realizing it. It is a hard habit to break and requires much conscious effort. This habit is more severe in the case of kids and more in school going kids.

The MIT Media Lab has come out with a technology called "Saving Face" [2]. It is a suite of easily scalable technologies to alert people when they are about to touch their faces and, in turn, help them to fight the pandemic.

Researchers from the Media Lab community came up with three classes of sensing technologies capable of detecting hand-face proximity and alerting the wearer. The device has two modes of operation, namely "record mode" and "alert mode." In "record" mode, users can train themselves to break the frequent face touch habit by learning how often they touch their faces, while "alert" mode will directly warn them with an alarm or vibration whenever they attempt to touch their face [2].

The Design Approach of the "Face Touch" Detection System Include

1. Using a SONAR-inspired approach to measure the distance between hands and face and warning the user with an alarm when they get too close. This technique transmits an ultrasound signal from earbuds and receives it with a microphone attached to the earphone near the face (neck). Many inexpensive off-the-shelf wired earbuds can generate and detect 20 kHz ultrasonic frequencies. In principle, users can use this approach with a smartphone for less than US $5 in headphones and cables. A mobile app offering both record and alert modes has been developed for the purpose. The open-source alpha version is available in Github (https://git hub.com/camilorq/SavingFaceApp) [2].

2. The second approach is based on the principle of magnetic flux change and detection. A device (https://github.com/irmandyw/magsense) has been developed that vibrates when the hand and face get too close. The device consists of a transmitter ring or bracelet fitted with a tiny magnet, which creates a magnetic flux. A receiving sensor (magnetometer), along with a controller embedded with the necessary code, is developed as a wearable device around the face as a necklace or a clip-on. Whenever the hand comes closer to the face, the magnetometer detects the flux using the electromagnetic or capacitive fields and generates an alarm in the form of beep sound or vibration.

3. The third approach is based on detecting the distance between a user's smartwatch and earbuds using the strength of the Bluetooth Low Energy (BLE) signal. The received signal strength indicator (RSSI) is measured while the communication between the Bluetooth module in the smartwatch and earbud takes place. The RSSI value indicates the approximate distance between two BLE modules, and this principle is used to find the distance between the earbud (close to the face) and the smartwatch (hand). When the distance is below a threshold, an alert can be generated as the user is approaching the face with their hand. This model can

also be used to count the number of face touches or how many times the user is attempting to face touch.

All the above methods can be sensitive enough to detect a hand 20 cm from the face, which is far enough to deliver a timely warning in "alert" mode. The required mobile applications are developed, which are compatible with off-the-shelf hardware for this application. These devices do not stop anybody from touching their faces but only alerts the user through beep sound or vibration. If the user needs to touch the face and is sure about the hand's cleanliness, they can ignore the alert and still touch their faces.

An Intelligent IoT Framework for Gesture Recognition to Achieve Contactless Interface and Social Distance [3, 4]

Control system design for controlling different devices through gesture recognition has been in practice for quite some time. Gesture recognition is an approach toward computing user interface that allows the system to capture and interpret human gestures as commands. One of the most intuitive methods for human-machine interaction is gesture spotting. These days, robots are performing predefined movements based on the gestures of human beings. These movements are generally executed in sequence to perform complex tasks. Gesture recognition can especially help persons with disabilities by having them wear sensors on their bodies.

Contactless interface is a key to control the spread of the COVID-19 infection among the people. Different devices required in our day-to-day activities like using lifts/ escalators, opening door locks, controlling sanitizer dispensing systems, sending commands to a computing system, etc. can be achieved through gestures instead of touches. Avoiding touch, in any case, is the need of the hour.

A gesture is posture or movement of the user's body, generally hand movements. Gestures can be used for interaction with others to convey messages and can convey messages through physical movements of the body, hand, face. Sometimes, facial expressions can also convey a message. Gestures also play an essential role in day-to-day life to interact with others. These gestures can also be used to communicate with a system known as gesture-based interaction.

Multimodal interfaces combining gesture, speech, and vision-based actions are expected to be available shortly for human-machine interaction. Natural ways of interaction with computers include speech-based communication wherein users directly communicate with the computer; gesture-based interaction where a human makes a specific predefined gesture to command a machine; and vision-based interaction where the computer does some predefined tasks concerning what it captures through the camera. These are natural ways of interaction that are more flexible and faster for interaction because speaking a word is more comfortable than typing through a keyboard. Speech-based interaction has the disadvantage that the same word may be pronounced differently by different people with different accents.

Speech recognition also suffers from external noise. Vision-based interaction has privacy problems, background effects, and limited coverage area.

Different types of sensors, along with various machine learning algorithms, have already been used for this type of application. Camera-based sensors and wearable-based sensors have been proved quite useful in this field. Although camera-based sensors provide higher accuracy, they suffer from high computational cost and are constrained by the field of view of cameras used in addition to privacy issues. Wearable sensors such as Inertial Measurement Unit (IMU), Electromyography (EMG) provide an energy-efficient and low computational cost in capturing gestures. EMG has been proved useful in health recovery monitoring, IMU in sign language recognition, and together they have been used to overcome limitations of each other. Despite their efficient performance, no popular and productive gesture-based control system has been developed and used.

In a gesture-driven interface, interaction with the controller happens using gestures of the human body, typically hand movements. One example of gesture-based interaction is that some mobile phones can take photos when we perform a predefined gesture. The gestures can be categorized as the following.

(a) Static gesture

In static gestures, users do not have to make any movement of the body while making a gesture. Statis gestures are used in our daily life to show digits using only fingers. In this case, there is no physical body movement; the body's position shows the predefined gesture. It is easier to classify static gestures compared to dynamic gestures.

The best application of static gestures is a sign language recognition system. In sign language, there is no physical movement of the body. The user simply shows characters of language by fingers as steady-state and position of fingers show particular gestures.

(b) Dynamic gesture

In dynamic gestures, the user makes body movement while making a gesture, e.g., waving hands to say goodbye. Here, action defines the gesture. Dynamic gestures are not more natural to classify because it includes lots of constraints like the direction of motion, speed of movement, and time elapsed for gesture presentation.

Both dynamic and static gestures are essential for system interaction and play a significant role in gesture-based systems for achieving touchless control. The touchless control can be achieved by using (1) ultrasonic sensor, (2) camera sensor, (3) accelerometer sensor, (4) gyro sensor among a few others. The complete system architecture is shown in Fig. 4.1.

In this gesture recognition system, the module records values from non-invasive body-mounted sensors, which are further transformed and processed by a model to extract features. This is accomplished by using methods of gesture segmentation and gesture recognition. The technique used for gesture segmentation is the adaptive

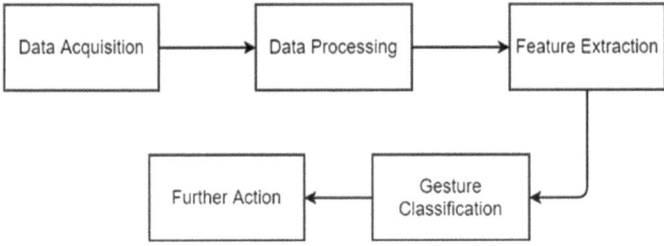

Fig. 4.1 Complete system architecture

threshold algorithm, and for gesture recognition, deep learning methods like convolutional neural network (CNN) will be used. Sensor Gyro 6050 is used to generate a dataset of digits. For each digit, a dataset of 300 samples has been created. ESP8266 controller delivers a highly integrated WiFi system on chip (SoC) solution, which is used as a controller in this work. It has efficient power usage, compact design, and reliable performance in the Internet of things framework. An accuracy of 96.9 has been achieved using the mentioned algorithm for detecting gestures (identification of digits).

Hardware and Connections Used

The primary purpose of hardware is to read the sensors' values corresponding to different gesture movements and send this as input to classification algorithms. The hardware consists of mainly two things, i.e., sensors,= and controller. The sensor is responsible for reading values corresponding to different gestures. Sensors like IMU, EMG play a vital role in gesture-based technology. The controller receives values from the sensor and does the further processing. The different hardwares used in this system are the following:

a. **ESP controller**

The controller receives reading values from the sensor, and further processing can be done. Here, it uses the ESP8266 controller, as shown in the following Fig. 4.2.

The ESP8266 is a low-cost system on chip (SoC) microchip. It has a TCP/IP stack to send data over the internet using WiFi networks. It is capable of hosting an application on it and can execute functions from another application processor over the WiFi network. It is mostly used in the Internet of things (IoT).

Features of ESP8266:
Low cost, compact
Supports 802.11 b/n/g WiFi protocol

Fig. 4.2 The NodeMCU
ESP8266 module [26]

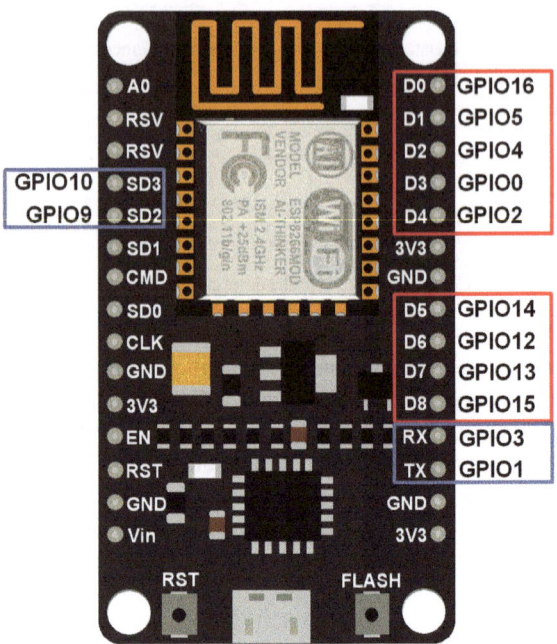

Power supply: 3.3v
Built-in low power 32-bit Tensilica
RISC processor
Current consumption: 100 mA
512kB flash memory
support deep sleep (40 μA)

Features: It can operate at input voltage within range 1.7–3.6 V. It comes with a Tensilica L106 32-bit RISC processor. This processor achieves low-power consumption and reaches a maximum clock speed of 160 MHz. It is also integrated with a real-time operating system (RTOS), which allows 80% of processing power to be available for user applications. ESP8266 module has the following features:

b. **MPU6050 (Concept of IMU)**

The system uses an IMU-based MPU6050 motion tracking device. MPU6050 is based on micro-electro-mechanical system (MEMS) technology. It can effectively collect information about gestures. It includes an accelerometer, gyroscope, and temperature sensor. The sensors have high sensitivity and can be fixed to wrist, arm, and other positions to obtain gesture data. It can be used to make new tools in gesture-based technology. This module is not affected by the external environment when collecting data and can be applied in a gesture-based interface. MPU6050

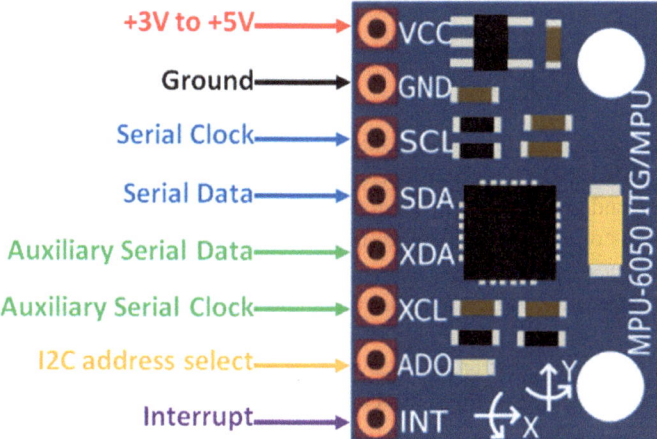

Fig. 4.3 The IMU6050 module with PIN description

can measure the force of body, angular rates, and sometimes the orientation of the body using a combination of three-axis accelerometers, three-axis gyroscopes, and at times magnetometers. The accelerometer is used for measuring linear acceleration and gyroscope for the rotational rate in 3D axes. One of the essential features of MPU6050 is that it has an in-built Digital Motion Processor (DMP), which is powerful enough to perform the computation required for sensor data fusion. It uses the I2C communication protocol for communication with peripheral devices. It needs an input power supply of 3–5 V. It is a low cost, low computational cost, not sensitive to the environment, and energy-constrained sensors. A total of six features are captured using a three-axis accelerometer (Accelx, Accely, Accelz) and three-axis gyroscope (Gyrox, Gyroy, Gyroz) sensors. Figure 4.3 shows the pin diagram of the unit.

Features of IMU6050:
Micro-electro-mechanical system [MEMS] technology.
Three-axis accelerometer and three-axis gyroscope.
16-bit ADC to get digitized output.
Power supply 3–5 v
Communication protocol I2C
DMP: Digital Motion Processor, which is powerful enough to perform the computation of sensor fusion.

Implementation

(a) **System architecture**:

The main goal is to design gesture-based interfaces to interact with computers. The system records values from non-invasive body-mounted sensors, which are further transformed and processed by a model to extract features. This is accomplished by using methods of gesture segmentation and gesture recognition. Gesture segmentation is the process of segmenting continuous data from sensor values to obtain actual data that contains a gesture using the threshold algorithm, and gesture recognition is the process of classifying this gesture into a predefined gesture. The following Fig. 4.4 shows the architecture of the system. Here, the controller continuously reads values from non-invasive body-mounted sensors. When sensor values reach the preset threshold value, system segments data are further transferred to classification algorithms. The classification algorithm classifies gestures into predefined clusters, and the system takes action according to the command given by gesture. The pictorial representation of the detailed process flow is given in Fig. 4.5.

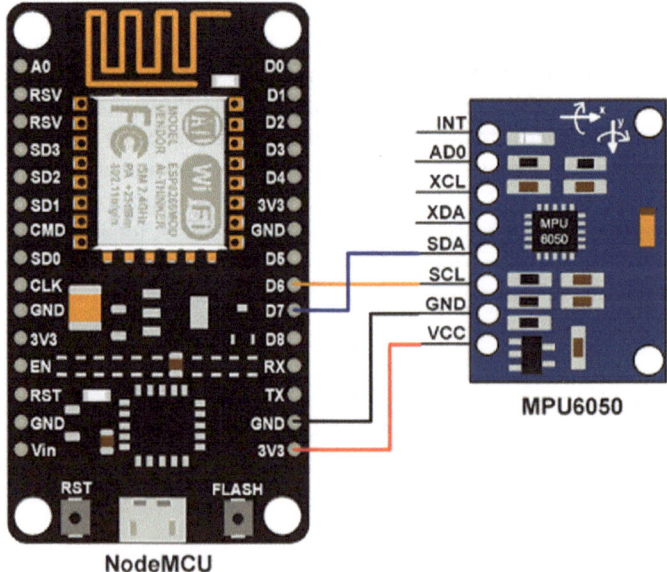

Fig. 4.4 Node MCU8266 and MPU6050 connection diagram

Fig. 4.5 System flowchart of the gesture recognition system

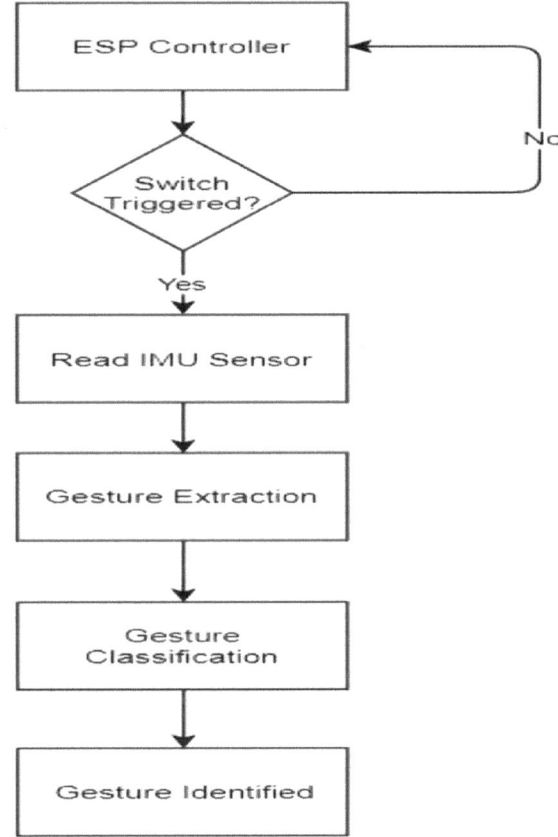

The Interface Between MPU6050 and ESP8266

Softwares used:
Arduino: IDE used to program ESP controllers.
ESP Flasher: Used to flash the code from ESP.
mPyloader: Used to transfer file and communication tool for ESP
Anaconda: used as programming IDE for CNN implementation

This research and design aim to achieve social distance, which is very important to restrict the spread of any virus, particularly the "Corona-Novid-19" virus, which is communicable and spreads through touch-sensitive human-to-human transmission. This idea presents techniques to achieve the following without any touch (which is a pivotal component to restrict the spread)

1. Opening of door lock through gesture (by communicating a secret series of digits (PIN or password) which will be communicated to a server and the server verifies the PIN and if authorized, opens the lock and door (achieves contactless home entry)
2. Controlling the lift without touching the buttons of the lift

    a. Without touching the lift button, a person can ask for the lift through gesture.
    b. Once inside the lift, a user can use the lift by communicating a digit using gestures.

3. Controlling the home appliances switches through gestures.
4. Virtual reality (VR):
   Virtual reality is also known as a virtual environment. Virtual reality is a computer simulation that can create and simulate virtual worlds. Equipment, such as VR, headset, and data gloves, allow users to sense and control various virtual objects in real time. It creates experiences that users cannot obtain in the real world and generating an actual response.
5. Data gloves: Data gloves are one of the essential devices for virtual reality gesture-based interaction. Data gloves can collect data of posture and motion of real-time human hand movement indicating gestures. Gloves contain different sensors mounted on it. Generally, it contains inertial movement unit (IMU) sensors. Data gloves can operate in two functions.
6. On the one hand, data gloves can act as an input device and collect the gesture movement of the user in real time and convert the received signal into virtual hand motion. Then, the user can observe the activity of the hand in virtual space through the movement of the virtual hand and can operate a virtual target by using various gestures. On the other hand, when using a feedback data glove in virtual space, the output of a feedback device enables the user to feel the target's physical property during operation and increase realism.

Smart Glasses for Monitoring and Controlling the Infection Spread [5]

This section describes the design, development, and workflow of the smart glasses for COVID-19 detection and follow-up action. The smart glass is equipped with two different cameras: (i) the optical camera to capture detected face and (ii) thermal camera to capture body temperature and then process this information for detecting COVID-19 infection. A thermal camera is utilized for hot body detection and recognition by adopting the high-temperature variability compared to other objects within the scanned area. This module uses image segmentation techniques on the recorded thermal image (heat map). It then superimposes the thermal image segments with the colored optical image to detect the object's face with high temperature. The face detection is processed using standard, and popular cascade classification algorithm [6], which is based on Haar features [7]. This system is completely developed using the Android platform for compatibility with smartphones. The GPS module determines the location coordinates after tagging it and stores it along with detected face and temperature so that other health officers can access the data through a smartphone. The medical officer and health administrators will get the data about the person with high temperature and identify people with suspected infection of COVID-19 through smart glasses, as shown in Fig. 4.6.

The system will notify the person concerned and the health administrator sending alert messages together with captured face and body temperature data of the suspected

Fig. 4.6 System flowchart of the smart glass system

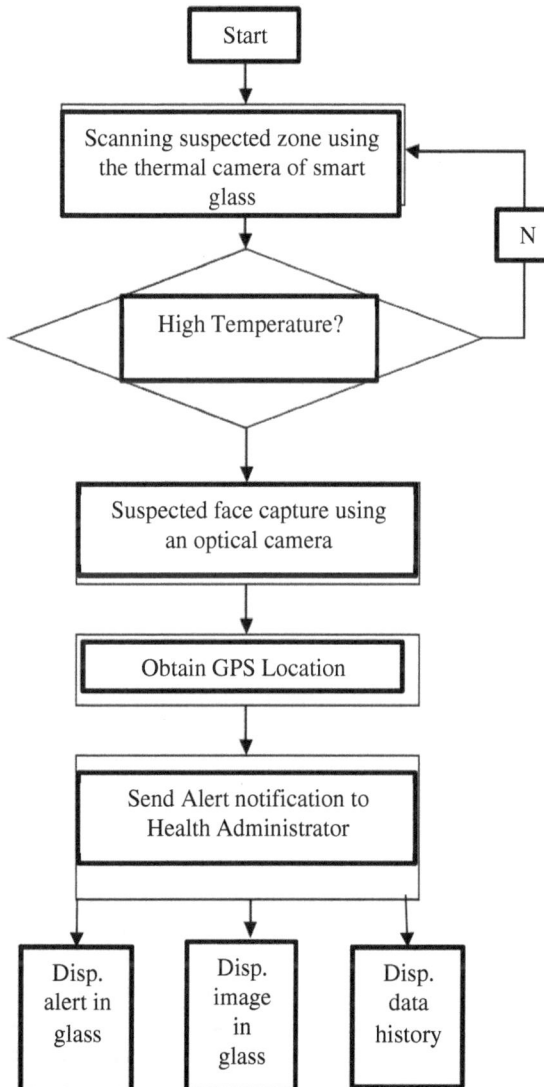

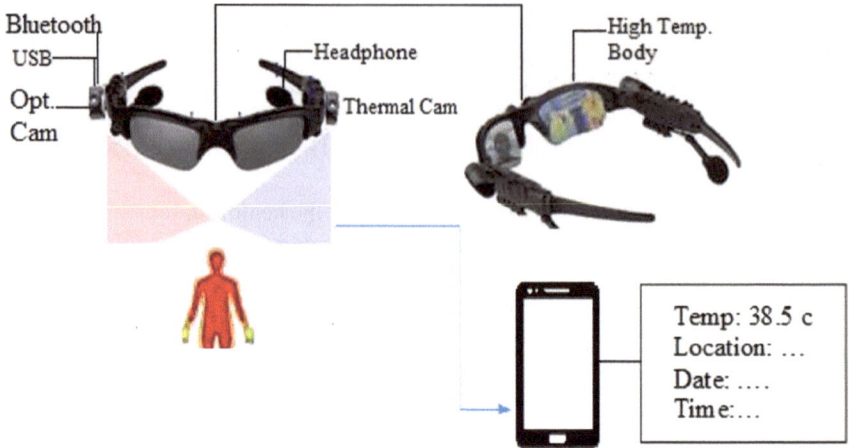

Fig. 4.7 Smart glass scanning and workflow diagram

individuals. The system is implemented using Android Studio and Java programming language. The Android Studio will generate the android application that can be run on the smart glass device. The thermal camera software development kit (SDK) also needs to be configured inside the system to capture thermal images. The Viola-Jones fast face detection method [8] can also be used for detecting the face from the input optical images.

The prototype's configuration is shown in Fig. 4.7, which describes the model with the thermal imaging, optical camera, Bluetooth communication, and Android framework for detection and identification of individuals with increased body temperature for the COVID-19 tracing and surveillance.

The Methodology

Thermography is the best method for scanning individuals and a large flow of people from a distant location in a contactless and non-intrusive manner. To do this, the temperature is measured, and an alarm triggered if it is above the standard threshold. This allows persons with increased body temperature to be identified quickly and reliably and isolated for more confirmatory testing. This smart glass-based system automatically detects high temperature via screening images, makes quicker diagnoses, and reduces human error risk, as shown in Fig. 4.7.

The screening process will be shown on the glasses screen that includes the detected face and the captured body temperature. The captured face and temperature are also stored in the database. The health officer and administrator can access the screening result after the data is shared from the smartphone. As soon as high body temperature is detected, the health officer will be notified through an alert message from the smart glasses. For example, when the thermal camera detects a high-temperature body, the system notifies the health officer to alert them regarding the risk. Besides, the system will take a picture using the optical camera and send it to the health officer.

All these are possible due to the rapid development in processing performance and memory, live video processing for computer vision in portable devices. The algorithms of conventional computer vision, image fusion, image segmentation, etc. have helped to develop the smart glasses system. Google location history (GLH) has been used to track user mobility, estimate the infection spread, control infectious diseases, and proper planning. The user's travel and contact history are connected to the user's Google account, like most Google features and services. The history of last visited places by the suspected carriers of the virus can be obtained through the Google location history (GLH) and can be used for contact tracing.

Smart Wearable Thermometer for Continuous Temperature Monitoring: [9, 10]

The researchers at IIT-Istituto Italiano di Tecnologia have developed a prototype sensor suit, namely "Smart Wearable Thermometer," that can quickly screen human body parameters [9, 10]. Also, the smart band ready clients are triggered when their internal (body) temperature level is higher than normal temperature. While collecting the vital parameters of the human body, the smart band also emits radio signals, which another armband can receive. When two smart bands are coming closer, they estimate the distance between them using the received signal strength indicator (RSSI) value. If the estimated distance is too close (less than two meters), the band generates alerts in the form of vibration. In this manner, transmitting an alarm signal is used to help individuals maintain the most important "social distance" norms. The transmission of the radio signal uses an ISM frequency of 2.45 GHz, equivalent to Bluetooth communication [9, 10].

In many countries, individuals use armband and ring-like devices powered with AI strategy and cloud interface through "CloudMinds" [11] to give consistent checking of indicative symptoms like temperature, pulse, and blood oxygen levels. Hong Kong is also utilizing electronic tracker wristbands to alarm healthcare specialists and government authorities for people having international travel history and does not agree to home isolation [9, 10].

Connected Thermometers for Large-Scale Temperature Screening

Connected (network) thermometers [12] are being utilized by emergency clinics to screen patients and medical staff at different emergency clinics all through China and a few other countries. At present, they use a network of temperature sensors to screen COVID-19 patients to decrease the dangers of infection spread among medical staff. This device is designed and developed by a California-based wellbeing startup VivaLNK's [12].

As shown in Fig. 4.8, this representative device uses a temperature sensor to monitor the temperature level continuously. This uses an IoT-based access controller to get ongoing patient information from the sensors and remotely transmit it to a healthcare station for continuous monitoring. This unit allows simultaneous data collection from a maximum of 40 Bluetooth Low Energy (BLE) devices to be combined and aggregated that is expected to cover numerous rooms in the clinic. Besides, from this, Kinsa Health [13] has utilized information collected from its more than one million associated thermometers to create day-by-day heat maps demonstrating which US areas are seeing an increase in detection and expansion of high fever.

Fig. 4.8 System to enable contactless control

Small and User-Friendly Contactless Thermometer

Temp Pal, [14], which is famous as the world's smallest and most efficient thermometer, offers a cloud-based nonstop remote framework for temperature measurement and alerts clients/ users whenever their body temperature is increased beyond an acceptable level. It is a small stamp-sized form factor with a weight of 3 grams and runs for 36 h for every battery recharge. It transmits temperature information through Bluetooth Low Energy to iWEECARE's versatile application cloud [14]. It supports a one-to-many transmission protocol and helps with timely clinical treatment. The Temp Pal Cloud Cluster System helps control the spread of COVID-19 by monitoring the individuals in self-isolation [14].

IoT Buttons for Generating Alert for Maintaining Cleanliness

Visionstate structured its first Internet of things buttons [15], which was first used in medical clinics in Vancouver, Canda. The buttons are called Wanda QuickTouch that are used to send different types of requirements of the building premises, laboratories, nursing stations to a central controlling station. As shown in Fig. 4.9, the IoT buttons send instant alarms to the supervisors/ administrators concerning any cleaning or upkeep activities to avoid the spread of infection.

The innovation empowers office supervisors to follow cautions and staff reaction times. The button interface to the central dashboard screen to remind the staff about the work to be carried out [15]. The IoT buttons are battery operated and equipped

Fig. 4.9 Smart IoT buttons

with the LTE-M communication interface and are available in persistent rooms, nursing stations, bathrooms, or necessary zones. The buttons can be utilized related to cleaning and upkeep exercises all through an office building/ premises [15].

Smart Helmet-Based Novel COVID-19 Detection and Diagnosis: [16]

A helmet-like device based on IoT architecture has been proposed to detect COVID-19 infection and issue alerts. The smart helmet was equipped with two different types of cameras, allowing the gathering of detailed information of the face detection details and also temperature measurements. Optical camera and thermal infrared camera, which measures the temperature in a non-contact manner, try to find objects of interest. A thermographic camera or thermal imaging camera is a device that uses infrared radiation to create an image similar to that of a traditional camera that utilizes visible light to produce an image. This module of the smart helmet follows a segmentation approach for the image using the recorded temperature and captured color images by both thermal and optical cameras. Thermal cameras are utilized for hot body detection (based on the temperature) by observing the variability of high temperature compared with other objects within the scanned zone. If a thermal camera visualizes high temperature, then it creates high-intensity levels of infrared spectra. In this project, the Arduino IDE (Arduino integrated development environment) is used, which is written in Java language and represents a cross-platform application. This Arduino IDE contains many features like code editor, syntax highlighting, auto-indentation, and brace matching among a few others. The IDE additionally uploaded to an Arduino board by compiled and uploaded programs using an essential one-click mechanism. It also supports C and C++ languages in using special rules.

Further, it utilizes a wiring project that can produce several input-output methods to provide a software library known as wiring. Besides, Proteus software includes schematic, simulation, and circuit design. It is mainly used for drawing several schematics and performing real-time circuit simulation that empowers users to get access during the running phase, and thus creating a real-time simulation. For the face detection process, this prototype uses the EmguCV cross-platform [17], net wrapper, Intel OpenCV image processing library, C#, and.Net framework. The standard APIs are generated during programming by the OpenCV library. The face detection is done using the cascade classification algorithm proposed by Paul Viola and Michael [18]. This algorithm mainly uses the Harr feature for face detection. This feature is quite useful in this project as during COVID-19, people are advised to wear masks and cover most parts of the face. Further, a machine learning algorithm is used with a cascade function to train both positive and negative images. The open CV library already has the cascade object detection that recognizes the face of the captured

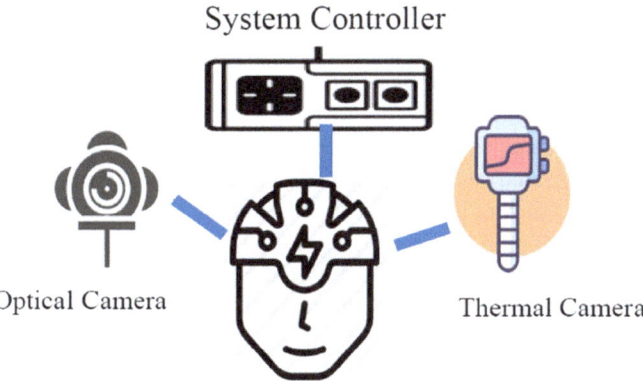

Fig. 4.10 Smart helmet design with the system controller, thermal camera, and optical camera

image. Many standard features are extracted from the human face to make a standardized size rectangle to enable standard image preprocessing algorithms to work on this. Figure 4.10 shows the smart helmet design model.

Through the smart helmet system, persons with increased body temperature can be identified quickly and reliably, and to be isolated for more exact testing and follow-up action. Beyond checking body temperature, artificial intelligence is being used to diagnose COV19. "Infervision" software platform that automatically detects symptoms via screening images and can make diagnoses quicker and reduce the risk of human error is also integrated with this system.

Figure 4.11 shows the detailed configuration of the system. The system concerns applicable thermal imaging frameworks for the detection and identification of an increase in body temperature as well as the surveillance process. This system uses an

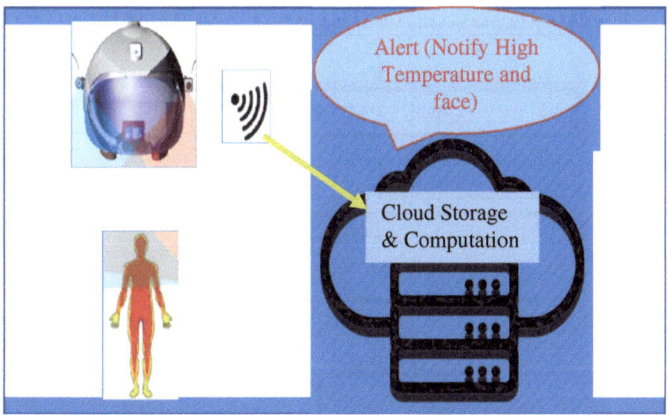

Fig. 4.11 System configuration and workflow of a smart helmet system [16]

enhanced helmet-fixable camera system that can be promptly deployed and utilized to visualize a thermal image with high resolution for the infected site.

The thermal imaging is coordinated to an optical camera for detecting the high temperature as well as the infected person in a contactless manner in a sensitive zone efficiently. Thermography is an ideal method for scanning individuals and large flows of people like at airports, railway stations, supermarkets, movie halls, and many such areas. For this, the temperature is measured, and alarm triggers in case it crosses the threshold value.

Face recognition has been exhaustively studied in the last few decades, and this device uses standard face recognition algorithms. Figure 4.12 illustrates the workflow of the proposed smart helmet.

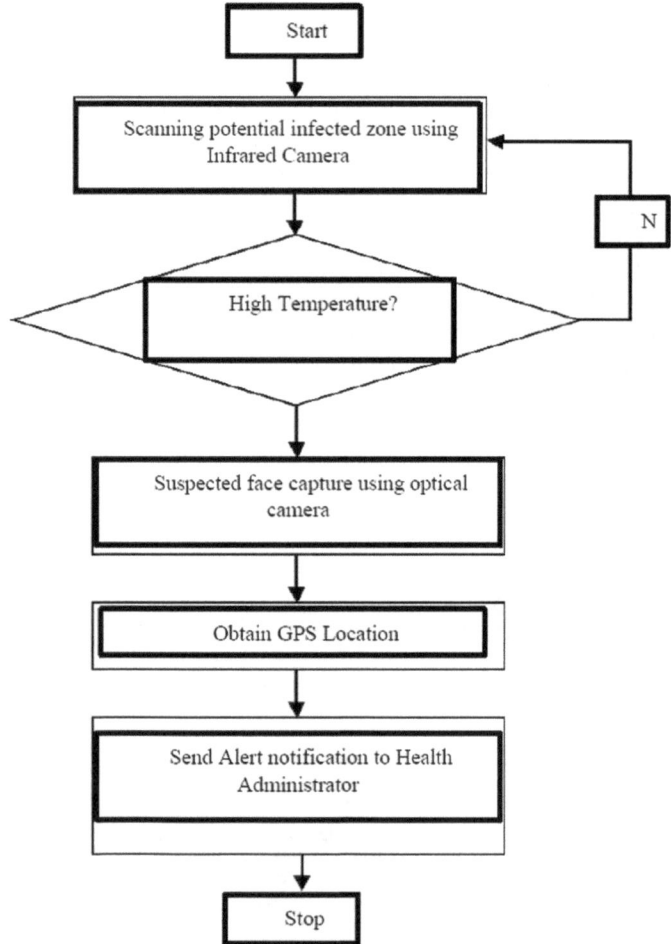

Fig. 4.12 Workflow description of smart helmet system

IoT-Based Drone Technology for Mitigating Challenges Owing to Coronavirus [19]

In this proposed system, an unmanned aerial vehicle (UAV) is used to gather more information about the COVID-19 infected people whenever it is required. A significant module in UAV is the drone control module. This system module is interfaced with a GSM module and IoT technologies, as shown in Fig. 4.13. This system is designed to deliver an emergency call to the central station in case it detects the temperature of a person more than a threshold (suffering from fever). The emergency message includes the location tag (measured by the attached GPS module) information of the person detected with high temperature. The entire emergency call is handled by the smartphone app working with the GSM mobile network. The UAV design consists of three parts. The first part details the mechanism of the input source. The processor development is carried out in the second part where the microcontroller is integrated, utilizing the Arduino IDE for embedded coding. The third part of the system is focused on the mechanism of output source, which consists of the integration of mechanical segments.

The UAV is equipped with two different cameras, allowing the gathering of detailed information on the temperature and the face of the people. Optical camera and thermal camera provide information about the temperature at which the different faces of interest are spotted. This module relates to an approach of image segmentation according to recorded temperature and captured colored images obtained by the thermal and optical cameras, respectively. A thermal camera is utilized for hot body detection and recognition by adopting the high-temperature variability compared

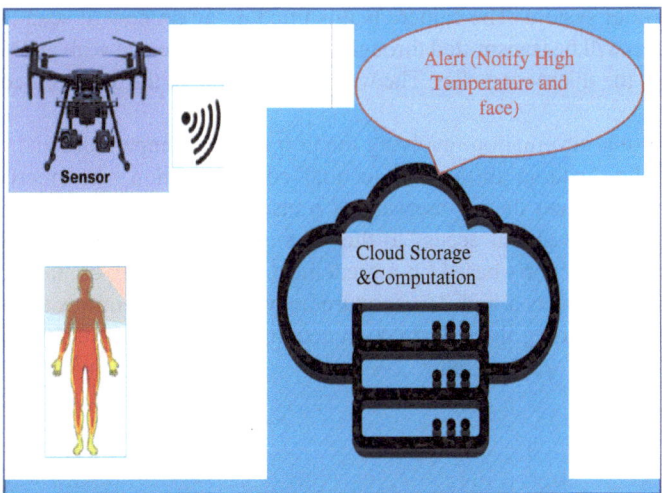

Fig. 4.13 Workflow of drone-based surveillance system

to other things within the scanned area. If the thermal camera visualizes the high-temperature body, it generates high-intensity levels of the infrared spectra. Since the pilot controls the drone to the designated area, virtual reality is being used along with live video monitoring to control the camera to scan people.

After the drone arrives at a specific location, it will start to scan the people's body temperature. The cameras that do the scanning will then transmit the live video monitoring clip to the smartphone. The live video monitoring is connected with an application on a smartphone so that the user can receive continuous live scanning from the entire flight by the drone. It will also create a realistic search operation during the flight but less interaction with people.

The drone control module is in charge of controlling both plans and actions of the drone. The drone is deployed from the central station after establishing the trajectory for its independent journey. The targeted destination point of the way was the GPS location gotten through the emergency call. The provided mechanism was tracked entirely and monitored within the control unit with the assistance of an online video stream. A GSM association and IP-based communication were coordinated to guarantee that there was a consistent connection between the central station and the field individual. The proposed system involves two cameras that track and monitor the assigned scanned zone. The drone is deployed to perform depicting visual and thermal cameras and collect more information for a precise hazard investigation. Optical and thermal sensors will be utilizing algorithms for detecting the hot body temperature by using an image processing module to distinguish and evaluate the dangers.

Drones and smartphones are integrated using WiFi signals, while camera control uses gyroscope data from an Android smartphone. The gyroscope is encompassed or embedded within the IMU. The drone IMU, along with GPS module, is used in the flight controller system. The camera that is fitted in the drone will scan the people; the live video will be transmitted through the headset connected with an IMU, which interfaces to the microcontroller. The working principle of the drone based is shown in Fig. 4.14.

After getting information involving the proper body temperature and GPS position from the Arduino through sequential communication, the microcontroller (NodeMCU) that had these observations transferred to it over the Web to allow autonomous online global access to this data. Moreover, the system collects and delivers situation reports based on a predefined schedule or on-demand. When the thermal camera detects a high-temperature body, the system notifies the authorities to warn them regarding the risk. In conjunction, the system will take a picture and sent it to the health administrator.

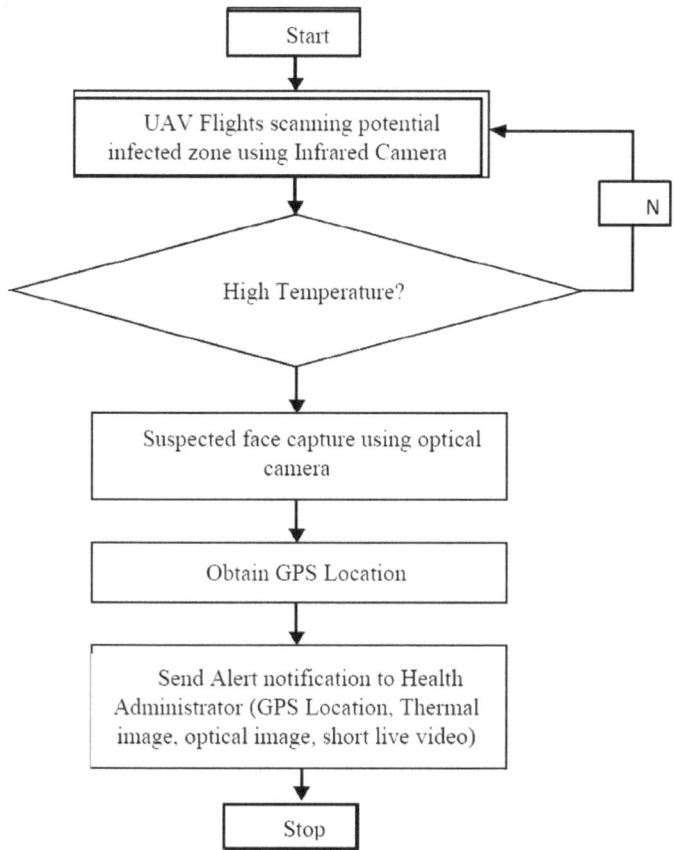

Fig. 4.14 Drone-based surveillance monitoring [19]

IoT-Based Wearable Band to Track COVID-19 Quarantine Patients: [20]

The tracing of COVID-19 positive patients is brought out generally through versatile applications, e.g., Aarogya Setu. As of April 21, 2020, Aarogya Setu takes a shot at the idea of publicly supporting. According to this current application's client audits, the usefulness of cautions to close by clients is restricted, since it accepts that each individual that has a cell phone has this application introduced. The clients have likewise demonstrated worries about the authenticity of the information entered in the application, proposing a requirement for checking or supporting power. Although the versatile application based on the following appears to be lacking, a visual marker-based following technique (e.g., clinical specialists stepping on the hands with a non-launderable ink) is best in recognizing the slipping off COVID-19 isolate subjects. Along these lines, to identify the quarantine subjects on schedule, none of the two referenced plans, as mentioned above, appears to work effectively and some way or another disregards the protection of subjects too. Different healthcare providers propose the utilization of wearable devices makes a patient increasingly consistent with clinical schedules and limitations; along these lines, a wearable band would be a correct answer for this case. Subsequently, a structured wearable band packaged with a versatile application would distinguish and follow the quarantine subjects continuously.

This wearable band is implied possibly to be taken off when a pre-decided isolate period is finished. Concerned clinical specialists have control of the following framework and answerable for the underlying subject enrollment, which lessens the security worries of the isolated subjects just as guarantees the authenticity of the information. GPS-based geofencing creates a framework for cautions and permits specialists to recognize isolated patients progressively. The versatile application additionally reports about any disembarkation or altering of the wearable band during a functioning isolate. Figure 4.15 displays the IoT-Q-Band framework's design. As the wearable band is fueled up with a battery and ought to be lightweight for agreeable wear, have re-utilized a few cell phone highlights, e.g., Web, and GPS get to utilize the application.

The IoT-Q-Band framework integrates clinical facilities and also liable for setting the term of the isolate and confirming the quarantine protocols. The wearable band is worn by an isolated subject either on the hand, arm, or leg and remotely associated with the portable application using a Bluetooth interface. The handling unit is the ESP32. After detecting, the band transmits the status (a byte of information) to the portable application at like clockwork stretch. The subject will be enlisted.

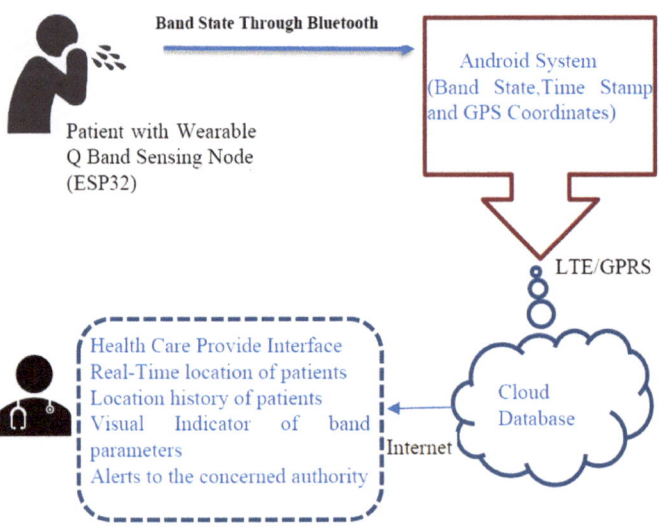

Fig. 4.15 Framework of IoT-Q-Band wearable system [20]

Fig. 4.16 IoT-Q-Band to trace the COVID-19 patient [20]

An assigned individual can screen each enrolled isolate case using a Web interface that brings every compelling case and present in a decipherable structure. The progression of the information beginning from the wearable band until the observing Web interface (administrator dashboard board). Figure 4.16 depicts the system embedded in the cloth and to be worn by the patient at the wrist and leg [20].

COVID-19 Intelligent Diagnosis and Treatment Assistant Program (NCapp) [21]

It is an application based on the Internet of things. Eight different modules are implemented in real-time online communication with the "cloud." All relevant and essential health parameters, together with travel history, possible contacts, are communicated to a designated cloud. According to these data, questionnaire responses, and check results, the diagnosis is automatically concluded as confirmed or suspected COVID-19 infection. It classifies patients into mild, moderate, or severe. The nCapp application also established an online COVID-19 real-time database, and it keeps updating the model of diagnosis in real time based on the latest real-world data to improve diagnostic accuracy.

Additionally, nCapp can act as a treatment support system. Physicians, experts, and managers are also linked to the nCapp platform for consultation. The nCapp system [21] also contributes to the long-term follow-up of patients who suffered from COVID-19 infection. The ultimate goal of the intelligent nCapp system is to enable different levels of COVID-19 diagnosis and treatment among different doctors from various hospitals that can be upgraded to the national and international standards. In this way, it can block disease transmission, avoid physician infection, and control the epidemic as soon as possible.

Robots and Drones in Service of the Society During COVID-19

Robots are being utilized to ease the weight on social distance and aid the treatment of patients. The best case of this is from the recently referenced field medical clinic in Wuhan, a joint endeavor between CloudMinds, China Mobile, and the Wuhan Wuchang Hospital [22]. At the Smart Field Hospital—5G-enabled robots to supply food, beverages, and drugs to patients. Notwithstanding, providing an essential break to staff and the robots' utilization restricted the social distance between infected patients. Robots from organizations like UVD Robots and Xenex Disinfection Services are being utilized to disinfect emergency clinics and others in affected areas in China, Italy, and the USA.

As discussed earlier, drones are being used to send clinical examples and supplies to and from COVID-19 hotspots [22]. The Japanese organization Terra Drone utilized drones to move supplies in China. Government authorities all through China, France, Spain, and the USA use drones to screen and guarantee delivery of essentials during the lockdown. Robots and drones are being utilized to splash sterilizing chemicals in some open spaces and on vehicles going between affected zones.

Systems That Indirectly Ensures Social Distancing

Different IoT-based systems are in use to ensure social distancing using Bluetooth communication, GPS-enabled systems, smartwatches, and other things. There are a few other systems that also indirectly helps in ensuring social distancing. Some of the following practices based on the sensors and IoT framework and other modes that indirectly help social distancing are as follows.

- Using existing CCTV and IP cameras to continually monitoring the movement of people and using video processing techniques to find the distance between them can ensure social distancing in many areas. The system can generate alerts whenever the computed distance between persons is less than the prescribed minimum distance of two meters.
- Contactless(touchless) participation framework: Many interesting innovations are now put to practice to promote touchless operations in day-to-day activities like gesture-based command control, foot operated doors, among few others.
- Telemedicine answer for healthcare institutions: Online and remote healthcare, which were there for quite some time, suddenly hog the limelight. People who were hesitant and questioning the efficacy of remote healthcare are now exploring this as a suitable alternative for the treatment of normal ailments. Some of the standard practices being followed are the following.

  – Specialist schedule planning
  – Conference utilizing video calling

Remedy dealing with the patient.

Monitoring Food Safety During the COVID-19 Pandemic [23]

The Internet of things (IoT) is changing the way restaurants and customers are engaged in the food business in the pre-COVID era. Today, both the restaurant owners and customers can trace products from the source to their doorstep of customers using IoT devices. The IoT device can monitor the delivery boy's location and the food's temperature while on the way. These devices have become very popular and quite helpful in keeping the food safe while in transit. This also helps in the proper management of inventory and giving owners real-time information to efficiently manage multiple locations.

Society is now more concerned about possible future outbreaks, and it can be a foodborne illness outbreak. The food industry should be more vigilant and alert to arrest any such outbreaks. Unfortunately, if it happens, we need to know which products are involved and the source location of the outbreak. IoT devices can be of great help in these scenarios as they allow owners to track their food from the time the order to the time it arrives. Even when the food items are on the highway, owners can still monitor the temperature and other physical and chemical parameters of the food, to ensure that safety standards are strictly followed.

Food-based businesses are now mandated to establish preventative control systems modeled after Hazard Analysis and Critical Control Points (HACCP) guidelines and prove their compliance by maintaining at least two years of data and documentation. IoT framework can help the business owners gather the data, store it for future auditing and, most importantly, ensure that the customers get the quality food.

Thus, utilizing the IoT for food safety is a critical aspect of quality control. These devices are equipped with a controller, temperature sensor, other required quality sensors like pH, dissolved oxygen, etc., barcode scanner, GPS module, and RFID-based infrared temperature reader to monitor and track the food throughout its journey in the supply chain.

- The temperature sensor, infrared camera, and RFID scanner track and measure each product's temperature during the supply chain.
- The IoT and cloud interface helps employees to complete checklists, including periodic temperature checks.
- The data collected in real time is immediately uploaded to a secure cloud that can be accessed anytime, from any location.
- Data analytics and user-friendly visualization can be done while the data is in the cloud. We can even customize, store, filter, and analyze the information according to our requirements.
- A detailed protocol for handling the food items can be prepared, and the sensors and IoT framework can be leveraged to ensure that. Alerts can be generated immediately if any steps of the protocol are overlooked, like non-observed items, missed

checklists, and corrective actions that address temperature or other parameter concerns.

- A detailed automated audit trail is generated and maintained to prove that it followed proper food safety protocol.

IoT Devices and Modern Dining Experiences [24]

In addition to the process of streamlining and managing day-to-day operations of the food business, IoT devices can also create a unique and pleasant dining experience for the customers. For example, if you love seafood, the restaurant can some restaurants can provide a complete track history (where and when seafood is harvested) using IoT devices. The "Boat-to-Plate" project funded by the Mid-Coast Fishermen's Association, developed a mobile app for the fisherman to upload information regarding their catch. A mobile app with IoT framework and sensors integrated is developed to find the quality and nutrient content of the food on the plate. The customers are more than happy to know the quality and the nutrient component of the food that is being served as they can avoid certain ingredients that are not good for their health. Restaurant owners these days are using IoT devices and related information to create many such unique dining experiences.

Inventory Control and Management [25]

IoT devices can efficiently manage and reduce the cost of inventory by providing real-time data that helps to optimally decide the time of ordering stock and forecasting requirements based on their demand. Tracking inventory from farm-to-fork has many advantages like preventing food waste, detecting in-house theft, and reducing the cost of managing the cost of inventory. Other questions and action items that IoT devices can help manage the inventory efficiently include:

- Who placed the order, who authorized the purchase, and when accepted the delivery?
- What is ordered, and what are the products' proper temperature and other parameter ranges?
- When did the order take place, and when did it arrive? When is its expiration date?
- What is the origin of the product, and how did it travel to reach the destination?

References

1. Ting, D.S.W., Carin, L., Dzau, V., et al.: Digital technology and COVID-19. Nat. Med. **26**, 459–461 (2020). https://doi.org/10.1038/s41591-020-0824-5
2. https://www.media.mit.edu/projects/saving-face/overview/
3. Ahmed, S., Cho, S.H.: Hand gesture recognition using an IR-UWB radar with an inception module-based classifier. Sensors **20**, 564 (2020)
4. Shanthakumar, V.A., Peng, C., Hansberger, J., et al.: Design and evaluation of a hand gesture recognition approach for real-time interactions. Multimedia Tools Appl. (2020). https://doi.org/10.1007/s11042-019-08520-1
5. Mohammed, N.A., Halimatuz Z., Al-Zubaidi, S., et al.: Novel COVID-19 detection and diagnosis system using IoT based smart helmet. Int. J. Psychosoc. Rehabil. **24**(7), 2296–2303 (2020). https://doi.org/10.37200/ijpr/v24i7/pr270221
6. Kotsiantis, S.B., Zaharakis, I.D., Pintelas, P.E.: Machine learning: a review of classification and combining techniques. Springer Science+Business Media B.V. (2007). https://doi.org/10.1007/s10462-007-9052-3
7. Hsieh, C., Liou, D.: Novel haar features for real-time hand gesture recognition using SVM. J. Real-Time Image Proc. **10**, 357–370 (2015). https://doi.org/10.1007/s11554-012-0295-0
8. Jones, M., Viola, P.: Fast multi-view face detection. Mitsubishi Electr. Res. Lab. TR2003-96 (2003). http://www.merl.com
9. https://www.mobihealthnews.com/news/asia-pacific/temp-pal-smart-thermometer-helps-reduce-covid-19-spread-hospitals
10. https://www.cnbc.com/2020/04/02/this-smart-thermometer-could-help-detect-covid-19-hotspots.html
11. Behar, J.A., Liu, C., Tsutsui, K., Corino, V.D.A., Singh, J., Pimentel, M.A.F., Karlen, W., Warrick, P., Zaunseder, S., Andreotti, F., Osipov, M., McSharry, P.E., Kotzen, K., Karmakar, C., Clifford, G.D.: Remote health monitoring in the time of COVID-19. 2005.08537, arXiv,cs.CY (2020)
12. http://www.vivalnk.com/covid-19
13. Chettri, S., Debnath, D., Devi, P.: Leveraging digital tools and technologies to alleviate COVID-19 pandemic. Available at SSRN: https://ssrn.com/abstract=3626092 or http://dx.doi.org/10.2139/ssrn.3626092 (2020)
14. https://www.iweecare.com/EN/index.html
15. https://www.visionstate.com/post/visionstate-ships-first-iot-buttons-for-rapid-response-to-cleaning-alerts
16. Mohammed, M.N., Syamsudin, H, Al-Zubaidi, S., Rusyaizila Ramli, S.A.K, Yusuf, E.: Novel COVID-19 detection and diagnosis system using IOT based smart helmet. https://doi.org/10.37200/ijpr/v24i7/pr270221
17. Gulzar, K., Sang, J., Tariq, O.: A cost effective method for automobile security based on detection and recognition of human face. In: 2017 2nd International Conference on Image, Vision and Computing (ICIVC), Chengdu, pp. 259–263 (2017). https://doi.org/10.1109/icivc.2017.7984557
18. Viola, P., Jones, M.J.: Robust real-time face detection. Int. J. Comput. Vision **57**, 137–154 (2004)
19. Mohammed, A., Hazairin, N., Al-Zubaidi, S., Karim, S., Mustapha, S., Yusuf, E.: Toward a novel design for coronavirus detection and diagnosis system using IoT based Drone Technology. Int. J. Psychosoc. Rehabil. **24**, 2287–2295 (2020). https://doi.org/10.37200/ijpr/v24i7/pr270220
20. Singh, V.K., Chandna, H., Kumar, A., Kumar, S., Upadhyay, N., Utkarsh, K.: IoT-Q-Band: a low cost internet of things based wearable band to detect and track absconding COVID-19 quarantine subjects. EAI Endorsed Trans. Internet Things
21. Bai, L., Yang, D., Wang, X., Tong, L., Zhu, X., Bai, C., et al.: Chinese experts' consensus on the Internet of things-aided diagnosis and treatment of coronavirus disease 2019. Clinical eHealth (2020). In press

22. https://www.cnbc.com/2020/04/03/covid-19-proves-the-need-for-social-robots-and-robot-ava tars-experts.html
23. https://www.marketsandmarkets.com/Market-Reports/covid-19-impact-on-iot-market-212 332561.html
24. https://foodsafetytech.com/column/iot-influences-restaurant-food-safety-management/
25. Ding, W.: Study of smart warehouse management system based on the IOT. In: Du, Z. (ed.) Intel- ligence Computation and Evolutionary Computation, pp. 203–207. Springer Berlin Heidelberg, Berlin, Heidelberg (2013). https://doi.org/10.1007/978-3-642-31656-2_30
26. http://www.nodemcu.com/index_en.html#fr_54747361d775ef1a3600000f

Chapter 5
Future Possible Applications

IoT is a creative innovation that guarantees that every single-tainted individual because of this COVID-19 infection is monitored effectively during the isolate. During isolation, it is useful for a real observing framework. All high-hazard patients are followed effectively utilizing the IoT system. This innovation is utilized for biometric estimations like circulatory strain, heartbeat, and glucose level. Figure 5.1 shows the primary benefits of IoT for COVID-19 pandemic.

With this innovation's fruitful usage, we can see an improvement in the effectiveness of clinical staff with a decrease in their remaining burden. The equivalent can be pertinent on account of COVID-19 pandemic with lesser costs and missteps.

Progressions Related to IoT for COVID-19

IoT is an inventive electro-mechanical stage to battle with COVID-19 pandemic and satisfy enormous difficulties during the lockdown circumstance. This innovation is useful to catch the constant information and other fundamental data of the tainted patient. Figure 5.2 shows the critical procedures utilized by IoT for COVID-19.

The first step involves the utilization of IoT to catch wellbeing information from different areas of the contaminated patient and deal with all the information utilizing the practical administration framework.

The significant effect of IoT in setting to COVID-19 concerns:

(i)   Contact tracing,
(ii) Infectious cluster determination,
(iii) Compliance and consistency of isolate.

As seen, the Internet of things idea uses the interconnected system for the successful stream of information exchange. It additionally empowers the social specialists, patients, regular citizens, and so on to be regarding the administration

© The Editor(s) (if applicable) and The Author(s), under exclusive license to Springer   83
Nature Singapore Pte Ltd. 2021
S. K. Udgata and N. K. Suryadevara, *Internet of Things and Sensor Network for COVID-19*, SpringerBriefs in Computational Intelligence,
https://doi.org/10.1007/978-981-15-7654-6_5

Fig. 5.1 Benefits of utilizing
IoT themes against
COVID-19 pandemic

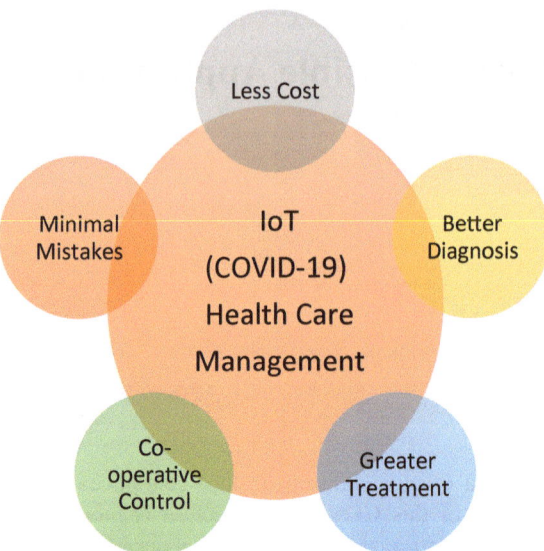

Fig. 5.2 Step process for
performing IoT tasks in
mitigating COVID-19

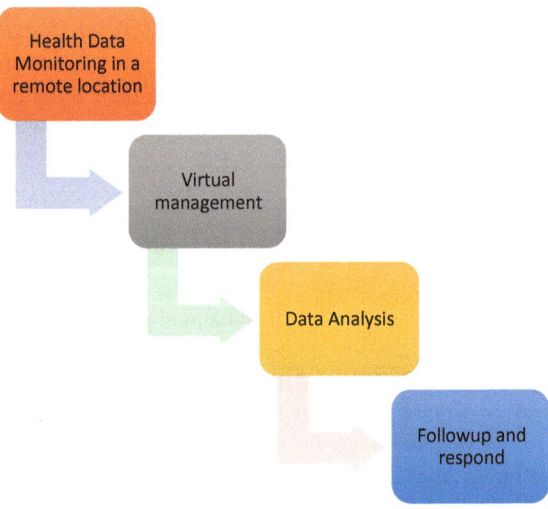

sponsors for talking about any issue and participation. Along these lines, by utilizing
the proposed IoT strategy in the COVID-19 pandemic, the compelling following of
the patient's care can be guaranteed. By building up a very much educated gathering regarding an associated organization, the group's recognizable proof can be
fundamentally made out. Some specific cell phone-based applications can likewise
be grown to the goal that the poor ones can receive profited in return.

Global Innovative Progressions to Determine COVID-19 Cases Quickly

Along these lines, to survive and make the regular citizens progressively mindful about the COVID-19 pandemic, India's administration has propelled a cell phone application named as, ArogyaSetu, which is meant to build up an association between the significant conceivable medicinal services administrations and the individuals of India. Correspondingly, in China, the portable application called, Close-Contact (English interpretation), is propelled for its regular citizens. This application informs the application holder concerning the closeness to the infectious individual, with the goal that the additional consideration can be taken while moving outside. After China, Taiwan immediately mobilized and organized explicit approaches for any conceivable coronavirus case recognizable proof, concealment, and asset arrangement to monitor the strength of the network. Taiwan gave and coordinated its national healthcare coverage database with its migration office and took the index to affect the production of sufficient information for examination; it created constant admonitions during a clinical visit dependent on movement relic and clinical side effects to help case distinguishing proof. They have additionally utilized this most recent innovation, which incorporates filtering of QR code, associated detailing of transport history, and so forth for the conceivable ID of the tainted ones.

Significant Uses of IoT for COVID-19 Pandemic

IoT utilizes countless interconnected gadgets to make a savvy organization for the best possible wellbeing of the individual's framework. It cautions and tracks any sorts of maladies to improve the security of the patient. It carefully catches the information and data of the patient with no human association. This information is additionally useful for proper dynamic procedure. Table 5.1 talks about the significant uses of IoT for COVID-19 pandemic. Technology helps to control the information and follow up on the report accomplished.

IoT is utilized for different applications to satisfy the significant prerequisite of lightening impacts of COVID-19 pandemic. The applications are applied for the appropriate administration of this pandemic. The patient can utilize IoT administrations for appropriate observing of pulse, circulatory strain, glucometer, and different exercises for customized consideration. It assists with following the wellbeing states of more seasoned individuals. The considerable utilization of this innovation in human services is to follow the continuous area of clinical hardware and gadgets for a smooth treatment process immediately. Medicinal services insurance agencies can utilize this innovation to identify misrepresentation guarantees and give straightforwardness in the general framework to improve the treatment work process of the patient with a practical exhibition and supportive of effective procedures during complex cases.

Table 5.1 Various use cases of IoT for COVID-19

Feature	Description
Internet of things healthcare units	The usage of IoT to help pandemic like COVID-19 needs a total incorporated system inside emergency clinic premises
Instant information to the healthcare provider throughout the crisis	This integrated system will permit the patients and the staff to react all the more rapidly and successfully at whatever point required
Fair COVID-19 treatment	The patients can profit the advantages offered by technology with no inclination of charges
Robotic treatment process	The determination of treatment techniques become gainful and helps the fitting treatment of the cases
Remote health monitoring and consultation	Makes the treatment accessible for the poor ones in the remote areas utilizing the all-around associated teleservices
Wireless system to identify and isolate COVID-19 patient	Various bonafide applications were introduced into cell phones, making the recognizable proof strategy smoother and progressively productive
Smart tracking of tainted patients	The significant tracing of patients reinforced the specialist organizations to deal with the cases all the more adroitly.
Real-time data collection and processing during the spread of this infection	As the gadgets, areas, channels, and so forth are all-around educated and associated, the on-time data sharing should be possible, and cases can be dealt with precisely
Swift COVID-19 screening	As the case showed up/found from the start occasion, the best possible analysis will have endeavored through associated treatment gadgets. Eventually makes the general screening process increasingly rapidly
Recognize novel solution	The general nature of management is the most ultimate objective. It very well may be accomplished by making advancements fruitful to the ground level
Connect every single clinical instrument and gadgets through the Internet	During COVID-19 treatment, IoT associated every clinical device and gadgets through the Web which pass on the constant data during treatment
Precise measuring of virus features	Based on the information report accessible, the utilization of some measurable strategy can likewise assist with foreseeing the circumstance in the next occasions. It will likewise assist with arranging the administration, specialists, and academicians to get ready for a superior workplace

Various Issues and Future Extent of the Investigation

The essential purpose of concern while utilizing the IoT in the current pandemic circumstance COVID-19 is about the security and protection of the information received, which is remarkable and essential from a quiet wellbeing perspective. The subsequent thing is about the consideration to be taken while coordinating the information arrangement among the gadgets in question and conventions. Figure 5.3 delineates the summed-up perspective on issues and difficulties in executing IoT for the COVID-19 pandemic.

It is irrefutable that innovation is getting us through the original estimates that have been set up because of the worldwide COVID-19 pandemic. Downloads and the utilization of working environment cooperation instruments, for example, Zoom, Microsoft Teams, and Slack, have expanded drastically and are permitting organizations to hold some similarities to arrange in the present exceptional condition. Simultaneously, Web-based life and video calling administrations like FaceTime permit families to remain associated despite complete physical disconnection in numerous areas. Along a comparative string, video spilling administrations give some type of diversion and a genuinely necessary break from the news.

Significantly, innovation is assuming a developing job in helping specialists forestall additionally spread of COVID-19, while likewise rewarding those that are shockingly contaminated. IoT, specifically and particularly when joined with other transformative advancements like cloud and AI, has seen use in a broad scope of uses during the emergency.

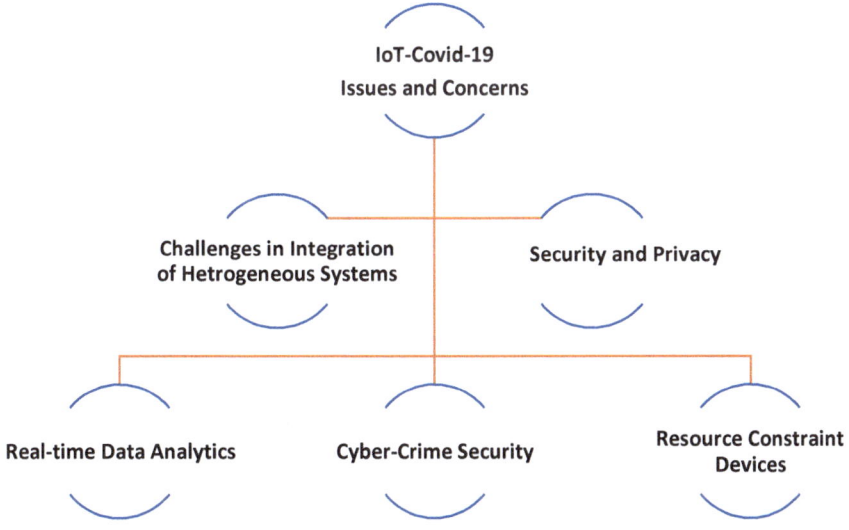

Fig. 5.3 Challenges and opportunities in realizing IoT for COVID-19

The universe of IoT is typically focused on the perception and enormous information of the things that our general public uses to support an association or industry to work. Models spread a wide range, for example, smart waste, smart parking, and in any event, building management. The adaptability is expanded exceptionally far. COVID has encouraged us all because if a computerized change methodology is not available, then we have to make them push ahead.

The potential test is that with numerous associations likely rising out of a financial downturn with fewer assets, they will require an approach to rapidly convey arrangements that assist them with tending to the real factors of a post-COVID-19 world.

Effect on the IoT Market Will be Blended

The conditions around COVID-19 will likewise, without a doubt, sway parts of the IoT advertise past medicinal services. A delayed period wherein "social separating" turns into the standard should bring about more noteworthy dependence of computerized arrangements in the scope of ventures.

Development limitations forced in numerous nations imply that a large number of representatives are telecommuting, and their standard work environments are closed. The period for arrival to typicality, including office work, is exceptionally dubious. The governments and organizations with physical offices are getting ready to come back to ordinariness. In the close to term, they will consider measures to confine the potential for a "second wave" of COVID-19. One of the potential patterns we may see is the move from biometric (finger- or thumb-based) staff participation/get to control frameworks to those dependent on touchless advances, for example, facial acknowledgment.

Many organizations are researching the methods of lessening the COVID-19 "surface region" in the workplace condition. The Singapore-based organization's development laboratory has discharged a facial acknowledgment-based time and participation framework that incorporates temperature recording by means of warm imaging innovation inserted into an entrance control screen and an IoT entryway. Generally speaking, the framework can likewise screen the development of high-temperature staff inside the workplace and send alarms to the board and HR. Be that as it may, while such use cases serve to help utilization of IoT, it is likewise evident that the huge monetary effect of the flare-up could constrain numerous associations to concede interest in new advances, for example, IoT. In a review of big business pioneers in 2019 done on the side of Digital Orbit Executive Briefing, 58% of respondents expressed that their associations were intending to submit critical assets to the selection of IoT arrangements, contrasted with just 4% that wanted to submit no or little assets. How COVID-19 affects assets distribution as it identifies with the reception of IoT and other transformative innovations will be a vital region of the focal point of examination for many organizations throughout the following year.

Specialized patterns inside the market could likewise be affected. For example, there is now a push to relocate investigation capacities from the cloud to the edge in some IoT applications, an arrangement that decreases inactivity and takes into account increasingly prompt reaction times. Edge computing likewise permits basic applications to work in any event when an organized network is down or debased. This pattern could be hurried by the utilization of IoT in a crucial application, such as checking the wellbeing vitals of travelers landing a plane.

The eagerly awaited take-up of 5G-which itself likewise conveys significantly better inertness—in IoT is, in all likelihood, set to be backed by different components. This remembers previously declared postponements for the endorsement of 5G gauges and systems organizations. On a progressively positive note, the episode of COVID-19 has not decreased mechanical organizations' enthusiasm for trialing/conveying private LTE/5G systems. The experience of the most recent couple of weeks has been that they should be adaptable and have become a standard working system for makers. This implies that adaptability over what items to fabricate, where and how to do it, and in what volumes. One of the significant drivers of 5G, close by adjoining gadgets and advances, for example, AGVs, machine vision, and 3D printing, will progress to adaptable creation. The experience of COVID-19 may well quicken this pattern.

IoT Technological Advancements for Healthcare Applications

Meshtastic

Meshtastic™ [1] is a task that lets you utilize reasonable GPS radios as extensible, long battery life, secure, work GPS communicator. These radios are connected to a typical activity where there is no dependable Internet/telecommunication setup. Every individual work can generally observe the area and send instant messages to other parts for a collaborative talk.

The Meshtastic radios naturally do work to advance bundles varying, so everybody in the gathering (group) can get messages from even the uttermost part. The radios will alternatively work with your telephone; however, no telephone is required. It can be used for open-air sports where cell inclusion is constrained (climbing, skiing, boating, paragliding, and gliders). It is mainly used for secure long-run correspondence inside gatherings (groups) without relying upon cell suppliers.

Through the Python API, applications can be realized using these cheap radios to effectively add work systems administration to your own undertakings. Some of the essential features are:

- Long battery life (ought to be around eight days with the beta programming)
- Worked in GPS and LoRa radio; however, we deal with the radio consequently

- Long range—a couple of miles for every hub except every hub will advance parcels varying
- Secure—AES256 encodes channels (but observe significant disclaimers underneath wrt this component)
- Shows course and separation to all individuals from the channel
- Coordinated or communicate instant messages for channel individuals
- Open and extensible codebase supporting various equipment sellers—no lock into one merchant.

Communication API can be used for Bluetooth gadgets (for example, Android application) to utilize the work. An iOS application is in progress. Furthermore, Meshtastic-Python gives access to work stations. Simply share an uncommon connection or QR code with companions, and they can join your scrambled work [1].

5G/6G Communications [2–4]

Every age of correspondence framework brings new and energizing highlights. The 5G correspondence framework, propelled worldwide in 2020, has energizing highlights. Be that as it may, 5G will not have the option to bolster the developing interest for remote correspondence in 2030 ultimately.

Exploration of 6G is still in its early stages and the examination stage. A few specialized issues should be tackled to send 6G correspondence frameworks effectively. A couple of potential concerns are quickly talked about beneath.

High spread and environmental retention of THz: The high THz frequencies give high information rates. Be that as it may, the THz groups need to beat a significant test for information move over generally significant distances on account of the high proliferation misfortune, and climatic ingestion qualities. This requires another plan for the handset design for the THz correspondence frameworks. The handset must be ready to work at high frequencies and have to guarantee the full utilization of broadly available bandwidths. In 6G, an exceptionally huge number of different sorts of correspondence frameworks, for example, recurrence groups, correspondence topologies, administration, and conveyance. will be included. Additionally, the entrance focuses, and versatile terminals will be mostly unique in the equipment settings. The monstrous MIMO procedure will be further updated from 5G to 6G, and this may require more complex engineering. The 6G framework will give full help to robotization frameworks, for example, self-governing vehicles, UAVs, and Industry 4.0 dependent on AI. To make self-sufficient remote frameworks, assembly of numerous heterogeneous sub-frameworks, for example, independent registering, interoperable procedures, the arrangement of frameworks, AI, self-ruling cloud, machines of frameworks, and different remote frameworks require adaptivity [2–4].

Smart Sensing: Improving the Capabilities of Sensors Based on AI Technologies

Computer-based intelligence and AI are the topics not so much comfortable by the investigators in the research areas of sensors and materials. There is a requirement to bridge the gap that can create applications using AI technologies for sensing systems. By this, the researchers will overcome any barrier that can make applications utilizing AI advancements for smart sensing.

As a matter of first importance, instrumentation strategies without artificial intelligence techniques cannot fulfill the incessant criterion of real-time sensing. The hardware cannot react to sensors' crisis demands. Second, the correspondence from sensors to various entities of the application is a significant issue, since sensors are with low data transmission and low vitality supplies. Third, as sensors are feeble in handling and correspondence capacities, it is hard for the application to ensure the security of the association or even endure blunders in that information. At last, when sensors transfer information to various things for storage and processing, how to guarantee the information security and protection condition is a significant issue.

The emphasizes the effect of AI on different sensor types, including optical, mechanical, and acoustic sensors for various applications, is to be investigated. The intricacies for the design and development of extra drivers using AI for smart edge processing can be realized so that new doors can be opened for future smart sensing and processing.

The IoT can possibly offer propelled types of assistance and applications across numerous spaces, and the energy that it has produced, along with its vast dreams, makes it a perfect wilderness for pushing electro-mechanical development. The empowering framework creates and scales, and the arrangement of gadgets turns out to be omnipresent.

References

1. https://github.com/meshtastic/Meshtastic-device
2. Chowdhury, M., Shahjalal, M., Ahmed, A., Jang, Y.: 6G wireless communication systems: applications requirements technologies challenges and research directions (2019)
3. Mumtaz, S., et al.: Terahertz communication for vehicular networks. IEEE Trans. Veh. Technol. **66**(7), 5617–5625 (2017)
4. Elliott, D., Keen, W., Miao, L.: Recent advances in connected and automated vehicles. J. Traffic Transp. Eng. **6**(2), 109–131 (2019)

Conclusion

IoT gives a broad, coordinated system to human services to battle with the COVID-19 pandemic. Every single clinical gadget is associated with the Web, and during any essential circumstance, it consequently passes on a message to the clinical staff. Contaminated cases can be taken care of properly in a remote area with all-around associated telegadgets. It handles all cases insightfully to offer at last fortified assistance to the patient and medicinal services. IoT is, by all accounts, a superb method to screen the contaminated patient. In medicinal services, this innovation is useful to keep up quality management with consistent data. By utilizing a factually based strategy, IoT gets supportive of anticipating an up and coming circumstance of this malady. With appropriate execution of this innovation, scientists, specialists, government, and academicians can make a better domain to battle with this infection.

Looking forward, the essential job of IoT as it identified with COVID-19 is probably going to be one of counteraction and assisting with identifying flare-ups before they arrive at mass scale. Ordered establishment of associated thermometers in air terminals could be an IoT application that before long gets typical. Another conceivable—however, likely longer-term usage—is a system of sensors that recognize hints of COVID-19. Upon recognition, an area could be "secured" to restrict spread and guarantee brief treatment to tainted people. It is not very difficult to envision such frameworks being fused into future city organizations, which as of now, incorporate applications planned for improving open wellbeing, for example, weapon discharge location and air quality sensors.

It ought to be brought up this hypothetical system of sensors that accumulates individual clinical information which could introduce common freedom questions and concerns. Indeed, even amid COVID-19 episode, WHO has communicated worries on this issue.

Security and protection concerns are not new to the IoT advertise. While challenges stay on this front, the issue is unquestionably one of expanded concentration for the whole business, which is adopting an increasingly all-encompassing strategy to point. In March 2020, Taoglas propelled the GDPR-consistent CROWD insights,

S. K. Udgata and N. K. Suryadevara, *Internet of Things and Sensor Network for COVID-19*, SpringerBriefs in Computational Intelligence, https://doi.org/10.1007/978-981-15-7654-6

a people development investigation stage that utilizes existing WiFi framework to measure, screen, foresee, alarm, and tell open get-together and social separating limit penetrate for indoor and outside scenes, for example, medical clinics.

Eventually, the COVID-19 flare-up—like 9/11—will probably change the analytics of what level of limitations and checking is worthy for everyone's benefit of society. The world has, without a doubt, transformed from the one that existed only weeks back. The monetary, social, and individual effect of COVID-19 is apparent and liable to bring about more noteworthy readiness from governments and general society to actualize frameworks that were close to home wellbeing data which is observed on a persistent premise.

The effect of IoT on the COVID-19 pandemic (and the other way around) has just started and is probably going to just develop in the up and coming months. While much is as yet obscure on how this circumstance will unfurl, obviously innovation is very much situated to help ventures, governments, and society that takes on the danger.